THE ECONOMICS OF GENOMIC MEDICINE

WORKSHOP SUMMARY

Adam C. Berger and Steve Olson, *Rapporteurs*

Roundtable on Translating Genomic-Based Research for Health

Board on Health Sciences Policy

INSTITUTE OF MEDICINE
OF THE NATIONAL ACADEMIES

THE NATIONAL ACADEMIES PRESS
Washington, D.C.
www.nap.edu

THE NATIONAL ACADEMIES PRESS   500 Fifth Street, NW   Washington, DC 20001

NOTICE: The project that is the subject of this report was approved by the Governing Board of the National Research Council, whose members are drawn from the councils of the National Academy of Sciences, the National Academy of Engineering, and the Institute of Medicine.

This project was supported by contracts between the National Academy of Sciences and the American Academy of Nursing (unnumbered contract); American College of Medical Genetics and Genomics (unnumbered contract); American Heart Association (unnumbered contract); American Medical Association (unnumbered contract); American Society of Human Genetics (unnumbered contract); Blue Cross and Blue Shield Association (unnumbered contract); Centers for Disease Control and Prevention (Contract No. 200-2011-38807); College of American Pathologists (unnumbered contract); Department of the Air Force (Contract No. FA7014-10-P-0072); Department of Veterans Affairs (Contract No. V101(93) P-2238); Eli Lilly and Company (Contract No. LRL-0028-07); Genetic Alliance (unnumbered contract); Health Resources and Services Administration (Contract No. HHSH250201100119P); Johnson & Johnson (unnumbered contract); The Kaiser Permanente Program Offices Community Benefit II at the East Bay Community Foundation (Contract No. 20121257); Life Technologies (unnumbered contract); National Cancer Institute (Contract No. N01-OD-4-2139, TO#189); National Coalition for Health Professional Education in Genetics (unnumbered contract); National Heart, Lung, and Blood Institute (Contract No. N01-OD-4-2139, TO#275); National Human Genome Research Institute (Contract No. N01-OD-4-2139, TO#264 and Contract No. HHSN263201200074I, TO#5); National Institute of Mental Health (Contract No. N01-OD-4-2139, TO#275); National Institute on Aging (Contract No. N01-OD-4-2139, TO#275); National Society of Genetic Counselors (unnumbered contract); Northrop Grumman Health IT (unnumbered contract); Office of Rare Diseases Research (Contract No. N01-OD-4-2139, TO#275); and Pfizer Inc. (Contract No. 140-N-1818071). Any opinions, findings, conclusions, or recommendations expressed in this publication are those of the authors and do not necessarily reflect the views of the organizations or agencies that provided support for the project.

International Standard Book Number-13: 978-0-309-26968-1
International Standard Book Number-10: 0-309-26968-7

Additional copies of this report are available for sale from the National Academies Press, 500 Fifth Street, NW, Keck 360, Washington, DC 20001; (800) 624-6242 or (202) 334-3313; http://www.nap.edu.

For more information about the Institute of Medicine, visit the IOM home page at: **www.iom.edu.**

Copyright 2013 by the National Academy of Sciences. All rights reserved.

Printed in the United States of America

The serpent has been a symbol of long life, healing, and knowledge among almost all cultures and religions since the beginning of recorded history. The serpent adopted as a logotype by the Institute of Medicine is a relief carving from ancient Greece, now held by the Staatliche Museen in Berlin.

Suggested citation: IOM (Institute of Medicine). 2013. *The economics of genomic medicine: Workshop summary.* Washington, DC: The National Academies Press.

*"Knowing is not enough; we must apply.
Willing is not enough; we must do."*
—Goethe

INSTITUTE OF MEDICINE
OF THE NATIONAL ACADEMIES

Advising the Nation. Improving Health.

THE NATIONAL ACADEMIES
Advisers to the Nation on Science, Engineering, and Medicine

The **National Academy of Sciences** is a private, nonprofit, self-perpetuating society of distinguished scholars engaged in scientific and engineering research, dedicated to the furtherance of science and technology and to their use for the general welfare. Upon the authority of the charter granted to it by the Congress in 1863, the Academy has a mandate that requires it to advise the federal government on scientific and technical matters. Dr. Ralph J. Cicerone is president of the National Academy of Sciences.

The **National Academy of Engineering** was established in 1964, under the charter of the National Academy of Sciences, as a parallel organization of outstanding engineers. It is autonomous in its administration and in the selection of its members, sharing with the National Academy of Sciences the responsibility for advising the federal government. The National Academy of Engineering also sponsors engineering programs aimed at meeting national needs, encourages education and research, and recognizes the superior achievements of engineers. Dr. Charles Vest is president of the National Academy of Engineering.

The **Institute of Medicine** was established in 1970 by the National Academy of Sciences to secure the services of eminent members of appropriate professions in the examination of policy matters pertaining to the health of the public. The Institute acts under the responsibility given to the National Academy of Sciences by its congressional charter to be an adviser to the federal government and, upon its own initiative, to identify issues of medical care, research, and education. Dr. Harvey V. Fineberg is president of the Institute of Medicine.

The **National Research Council** was organized by the National Academy of Sciences in 1916 to associate the broad community of science and technology with the Academy's purposes of furthering knowledge and advising the federal government. Functioning in accordance with general policies determined by the Academy, the Council has become the principal operating agency of both the National Academy of Sciences and the National Academy of Engineering in providing services to the government, the public, and the scientific and engineering communities. The Council is administered jointly by both Academies and the Institute of Medicine. Dr. Ralph J. Cicerone and Dr. Charles Vest are chair and vice chair, respectively, of the National Research Council.

www.national-academies.org

PLANNING COMMITTEE[1]

W. GREGORY FEERO (*Chair*), Contributing Editor, *Journal of the American Medical Association*, Chicago, IL
PAUL R. BILLINGS, Chief Medical Officer, Life Technologies, Carlsbad, CA
BRUCE BLUMBERG, Institutional Director of Graduate Medical Education, Northern California Kaiser Permanente, The Permanente Medical Group, Oakland, CA
DENISE E. BONDS, Medical Officer, Division of Prevention and Population Sciences, National Heart, Lung, and Blood Institute, Bethesda, MD
SARA COPELAND, Acting Chief, Genetic Services Branch, Health Resources and Services Administration, Rockville, MD
MOHAMED KHAN, Leader of Radiation Oncology, Vancouver Cancer Centre, BC Cancer Agency, Vancouver, BC, Canada
MUIN KHOURY, Director, National Office of Public Health Genomics, Centers for Disease Control and Prevention, Atlanta, GA
DEBRA LEONARD, Professor and Vice Chair for Laboratory Medicine and Director of the Clinical Laboratories, Weill Cornell Medical Center of Cornell University, New York, NY
MICHELE A. LLOYD-PURYEAR, Senior Medical and Scientific Advisor, National Institute of Child Health and Human Development, Bethesda, MD
JOAN A. SCOTT, Executive Director, National Coalition for Health Professional Education in Genetics, Lutherville, MD
KATHERINE JOHANSEN TABER, Senior Scientist, Genetics and Molecular Medicine, American Medical Association, Chicago, IL
MICHAEL S. WATSON, Executive Director, American College of Medical Genetics and Genomics, Bethesda, MD
CATHERINE A. WICKLUND, Past President, National Society of Genetic Counselors; Director, Graduate Program in Genetic Counseling; Associate Professor, Department of Obstetrics and Gynecology, Northwestern University, Chicago, IL

IOM Staff

ADAM C. BERGER, Project Director
CLAIRE F. GIAMMARIA, Research Associate (*until July 2012*)
TONIA E. DICKERSON, Senior Program Assistant

---

[1] Institute of Medicine planning committees are solely responsible for organizing the workshop, identifying topics, and choosing speakers. The responsibility for the published workshop summary rests with the workshop rapporteurs and the institution.

ROUNDTABLE ON TRANSLATING GENOMIC-BASED RESEARCH FOR HEALTH[1]

WYLIE BURKE (*Co-Chair*), Professor and Chair, Department of Bioethics and Humanities, University of Washington, Seattle

SHARON TERRY (*Co-Chair*), President and Chief Executive Officer, Genetic Alliance, Washington, DC

NAOMI ARONSON, Executive Director, Technology Evaluation Center, Blue Cross and Blue Shield Association, Chicago, IL

EUAN ANGUS ASHLEY, Representative of the American Heart Association; Director, Center for Inherited Cardiovascular Disease, Stanford University School of Medicine, Palo Alto, CA

PAUL R. BILLINGS, Chief Medical Officer, Life Technologies, Carlsbad, CA

BRUCE BLUMBERG, Institutional Director of Graduate Medical Education, Northern California Kaiser Permanente, The Permanente Medical Group, Oakland, CA

DENISE E. BONDS, Medical Officer, Division of Prevention and Population Sciences, National Heart, Lung, and Blood Institute, Bethesda, MD

PAMELA BRADLEY, Staff Fellow, Personalized Medicine Staff, Office of In Vitro Diagnostics and Radiological Health, Center for Devices and Radiological Health, U.S. Food and Drug Administration, Silver Spring, MD

PHILIP J. BROOKS, Health Scientist Administrator, Office of Rare Diseases Research, National Center for Advancing Translational Sciences, Rockville, MD

ANN CASHION, Acting Scientific Director, National Institute of Nursing Research, Bethesda, MD

C. THOMAS CASKEY, Professor, Baylor College of Medicine, Houston, TX

MICHAEL J. DOUGHERTY, Director of Education, American Society of Human Genetics, Bethesda, MD

VICTOR DZAU, President and Chief Executive Officer, Duke University Health System; Chancellor for Health Affairs, Duke University, Durham, NC

W. GREGORY FEERO, Contributing Editor, *Journal of the American Medical Association*, Chicago, IL

---

[1] Institute of Medicine forums and roundtables do not issue, review, or approve individual documents. The responsibility for the published workshop summary rests with the workshop rapporteurs and the institution.

ANDREW N. FREEDMAN, Branch Chief, Clinical and Translational Epidemiology Branch, Epidemiology and Genetics Research Program, Division of Cancer Control and Population Sciences, National Cancer Institute, Rockville, MD

GEOFFREY GINSBURG, Director, Center for Genomic Medicine, Institute for Genomic Sciences and Policy, Duke University, Durham, NC

RICHARD J. HODES, Director, National Institute on Aging, Bethesda, MD

SHARON KARDIA, Professor and Chair of Epidemiology; Director, Public Health Genetics Program; Director, Life Science and Society Program; Codirector, Center for Public Health and Community Genomics, University of Michigan School of Public Health, Ann Arbor

MOHAMED KHAN, Representative of the American Medical Association; Leader of Radiation Oncology, Vancouver Cancer Centre, BC Cancer Agency, Vancouver, BC, Canada

MUIN KHOURY, Director, National Office of Public Health Genomics, Centers for Disease Control and Prevention, Atlanta, GA

GABRIELA LAVEZZARI, Assistant Vice President, Scientific Affairs, PhRMA, Washington, DC

THOMAS LEHNER, Director, Office of Genomics Research Coordination, National Institute of Mental Health, Bethesda, MD

DEBRA LEONARD, Representative of the College of American Pathologists; Professor and Vice Chair for Laboratory Medicine and Director of the Clinical Laboratories, Weill Cornell Medical Center of Cornell University, New York, NY

ELIZABETH MANSFIELD, Director of the Personalized Medicine Staff, Office of In Vitro Diagnostics and Radiological Health, Center for Devices and Radiological Health, U.S. Food and Drug Administration, Silver Spring, MD

KATHRYN McLAUGHLIN, Public Health Analyst and Program Officer, Genetic Services Branch, Maternal and Child Health Bureau, Health Resources and Services Administration, Rockville, MD

KELLY McVEARRY, Senior Scientific Advisor, Information Systems Division, Northrop Grumman Health IT, Rockville, MD

ROBERT L. NUSSBAUM, Chief, Division of Medical Genetics, Department of Medicine and Institute of Human Genetics, University of California, San Francisco, School of Medicine

MICHELLE A. PENNY, Senior Director, Translational Medicine Group, Eli Lilly and Company, Indianapolis, IN

AIDAN POWER, Vice President and Head PharmaTx Precision Medicine, Pfizer Inc., Groton, CT

VICTORIA M. PRATT, Chief Director, Molecular Genetics, Quest Diagnostics Nichols Institute, Chantilly, VA

RONALD PRZYGODZKI, Associate Director for Genomic Medicine and Acting Director of Biomedical Laboratory Research and Development, Department of Veterans Affairs, Washington, DC

ALLEN D. ROSES, President and Chief Operating Officer, Cabernet, Shiraz and Zinfandel Pharmaceuticals; and Jefferson–Pilot Professor of Neurobiology and Genetics, Professor of Medicine (Neurology); Director, Deane Drug Discovery Institute; Senior Scholar, Fuqua School of Business, R. David Thomas Executive Training Center, Duke University, Durham, NC

KEVIN A. SCHULMAN, Professor of Medicine and Business Administration; Director, Center for Clinical and Genetic Economics; Associate Director, Duke Clinical Research Institute, Duke University School of Medicine, Durham, NC

JOAN A. SCOTT, Executive Director, National Coalition for Health Professional Education in Genetics, Lutherville, MD

DAVID VEENSTRA, Professor, Pharmaceutical Outcomes Research and Policy Program, Department of Pharmacy, University of Washington, Seattle

MICHAEL S. WATSON, Executive Director, American College of Medical Genetics and Genomics, Bethesda, MD

DANIEL WATTENDORF, Deputy Chief, Medical Innovations, Department of the Air Force; Program Manager, DARPA/Defense Sciences Office, Arlington, VA

CATHERINE A. WICKLUND, Past President, National Society of Genetic Counselors; Director, Graduate Program in Genetic Counseling; Associate Professor, Department of Obstetrics and Gynecology, Northwestern University, Chicago, IL

JANET WILLIAMS, Representative of the American Academy of Nursing; Professor of Nursing, The University of Iowa College of Nursing, Iowa City

Fellows

SEAN P. DAVID, James C. Puffer, M.D./American Board of Family Medicine Fellow

SAMUEL G. JOHNSON, American Association of Colleges of Pharmacy/American College of Clinical Pharmacy Anniversary Fellow

IOM Staff

ADAM C. BERGER, Project Director
CLAIRE F. GIAMMARIA, Research Associate (*until July 2012*)
TONIA E. DICKERSON, Senior Program Assistant

Board on Health Sciences Policy Staff

DONNA RANDALL, Administrative Assistant
ANDREW POPE, Director

Reviewers

This workshop summary has been reviewed in draft form by individuals chosen for their diverse perspectives and technical expertise, in accordance with procedures approved by the National Research Council's Report Review Committee. The purpose of this independent review is to provide candid and critical comments that will assist the institution in making its published workshop summary as sound as possible and to ensure that the workshop summary meets institutional standards for objectivity, evidence, and responsiveness to the study charge. The review comments and draft manuscript remain confidential to protect the integrity of the process. We wish to thank the following individuals for their review of this workshop summary.

James P. Evans, Department of Genetics, University of North Carolina at Chapel Hill
Deborah Heine, Claire Altman Heine Foundation, Inc.
David O. Meltzer, Section of Hospital Medicine and Center for Health and the Social Sciences, The University of Chicago
Scott Ramsey, Cancer Prevention Program, Division of Public Health Science, Fred Hutchinson Cancer Research Center

Although the reviewers listed above have provided many constructive comments and suggestions, they did not see the final draft of the workshop summary before its release. The review of this workshop summary was overseen by **Melvin Worth.** Appointed by the Institute of Medicine, he was responsible for making certain that an independent examination of this workshop summary was carried out in accordance with institutional procedures and that all review comments were carefully considered. Responsibility for the final content of this workshop summary rests entirely with the rapporteurs and the institution.

Acknowledgments

The support of the sponsors of the Institute of Medicine Roundtable on Translating Genomic-Based Research for Health was crucial to the planning and conduct of the workshop Assessing the Economics of Genomic Medicine and the development of the workshop summary report titled *The Economics of Genomic Medicine*. Federal sponsors are the Centers for Disease Control and Prevention; Department of the Air Force; Department of Veterans Affairs; Health Resources and Services Administration; National Cancer Institute; National Heart, Lung, and Blood Institute; National Human Genome Research Institute; National Institute of Mental Health; National Institute on Aging; and Office of Rare Diseases Research. Nonfederal sponsorship was provided by the American Academy of Nursing; American College of Medical Genetics and Genomics; American Heart Association; American Medical Association; American Society of Human Genetics; Blue Cross and Blue Shield Association; College of American Pathologists; Eli Lilly and Company; Genetic Alliance; Johnson & Johnson; The Kaiser Permanente Program Offices Community Benefit II at the East Bay Community Foundation; Life Technologies; National Coalition for Health Professional Education in Genetics; National Society of Genetic Counselors; Northrop Grumman Health IT; and Pfizer Inc.

The Roundtable wishes to express its gratitude to the expert speakers whose presentations helped outline the challenges and proposed potential solutions for assessing the economics of genomic medicine. The Roundtable also wishes to thank the members of the planning committee for their work in developing an outstanding workshop agenda. The project director would like to thank project staff who worked diligently to develop both the workshop and the resulting summary.

Contents

ABBREVIATIONS AND ACRONYMS   xix

1   INTRODUCTION AND OVERVIEW   1
      Organization of the Workshop, 2
      Major Themes of the Workshop, 4

2   GENOMICS, POPULATION HEALTH, AND TECHNOLOGY   9
      The Value of Genomic Data, 10
      The Long-Term and Mid-Term Promises of Genomics, 10
      Genomic Data in Healthy People, 11
      Challenges to Implementation, 13
      Another Medical Test, 14

3   THE INTERSECTION OF GENOMICS AND HEALTH
     ECONOMICS   15
      Economic Evaluation Tools, 16
      Incremental Cost-Effectiveness Ratios, 18
      Genome Sequencing, 18
      Comparative-Effectiveness Research, 20
      Three Challenges, 21

4   PRECONCEPTION CARE AND SEQUENCING   23
      A Clinician's Perspective, 24
      A Futurist's Perspective, 28

        A Patient's Perspective, 29
        Economic Perspectives, 32
        Discussion, 34

5   UNPROVOKED DEEP VEIN THROMBOSIS   37
        A Clinician's Perspective, 38
        A Futurist's Perspective, 41
        A Patient's Perspective, 42
        Economic Perspectives, 44
        Discussion, 47

6   CANCER CARE   49
        A Clinician's Perspective, 50
        A Futurist's Perspective, 51
        A Patient's Perspective, 52
        Economic Perspectives, 54
        Discussion, 55

7   PANELISTS' AND STAKEHOLDERS' PERSPECTIVES   59
        A Clinician's Perspective, 60
        A Researcher's Perspective, 61
        A Chief Scientific Officer's Perspective, 61
        A Patient's Perspective, 63
        A Public Health Officer's Perspective, 64
        A Hospital Administrator's Perspective, 65
        Economic Perspectives, 67
        Additional Issues, 70
        Closing Remarks, 70

REFERENCES   73

APPENDIXES

A   WORKSHOP AGENDA   77
B   SPEAKER BIOGRAPHICAL SKETCHES   87
C   STATEMENT OF TASK   99
D   REGISTERED ATTENDEES   101

Figure, Tables, and Box

FIGURE

3-1 The change in costs and change in effectiveness compared with current practice divides the results of cost-effectiveness analyses into four quadrants, 19

TABLES

3-1 Types of Economic Evaluations in Health Care, 17
3-2 Factors That Influence the Cost-Effectiveness of Genomic Testing Strategies, 20

BOX

7-1 Research Needs Identified by Individual Speakers, 71

Abbreviations and Acronyms

ACOG	American College of Obstetricians and Gynecologists
CEA	cost-effectiveness analysis
CLIA	Clinical Laboratory Improvement Amendments
CMS	Centers for Medicare & Medicaid Services
CUA	cost-utility analysis
EGFR	epidermal growth factor receptor
FDA	U.S. Food and Drug Administration
INR	international normalized ratio
IOM	Institute of Medicine
NIH	National Institutes of Health
QALY	quality-adjusted life year

1

Introduction and Overview[1]

The sequencing of the human genome and the identification of links between specific genetic variants and diseases have led to tremendous excitement over the potential of genomics to direct patient treatment toward more effective or less harmful interventions. Still, the use of whole genome sequencing challenges the traditional model of medical care where a test is ordered only when there is a clear indication for its use and a path for downstream clinical action is known. This has created a tension between experts who contend that using this information is premature and those who believe that having such information will empower health care providers and patients to make proactive decisions regarding lifestyle and treatment options. In addition, some stakeholders are concerned that genomic technologies will add costs to the health care system without providing commensurate benefits, and others think that health care costs could be reduced by identifying unnecessary or ineffective treatments.

Economic models are frequently used to anticipate the costs and benefits of new health care technologies, policies, and regulations. Economic studies also have been used to examine much more specific issues, such as comparing the outcomes and cost-effectiveness of two different drug treatments for the same condition. These kinds of analyses offer more than just

---

[1] The planning committee's role was limited to planning the workshop, and the workshop summary has been prepared by the workshop rapporteurs as a factual summary of what occurred at the workshop. Statements, recommendations, and opinions expressed are those of individual presenters and participants and are not necessarily endorsed or verified by the Institute of Medicine, and they should not be construed as reflecting any group consensus.

predictions of future health care costs. They provide information that is valuable when implementing and using new technologies. Unfortunately, however, these economic assessments are often limited by a lack of data on which to base the examination. This particularly affects health economics, which includes many factors for which current methods are inadequate for assessing, such as personal utility, social utility, and patient preference.

To understand better the health economic issues that may arise in the course of integrating genomic data into health care, the Roundtable on Translating Genomic-Based Research for Health hosted a workshop in Washington, DC, on July 17–18, 2012, that brought together economists, regulators, payers, biomedical researchers, patients, providers, and other stakeholders to discuss the many factors that may influence this implementation. The workshop was one of a series that the roundtable has held on this topic, but it was the first focused specifically on economic issues.

ORGANIZATION OF THE WORKSHOP

To have a focused discussion on the potential downstream health economic issues that arise from various models of using whole genome sequencing in clinical settings, participants were asked to make three assumptions: (1) whole genome sequencing costs are an acceptable and fixed expense, though interpretation costs may not be; (2) data storage costs are assumed to be acceptable and fixed as well; however, electronically stored data may not be transportable across health care systems over an individual's lifespan; and (3) such tests are available in the context of a health care encounter.

The workshop began with two broad overviews of the economics of genomic applications in medicine, the first from the perspective of a clinician (Chapter 2), and the second from the perspective of an economist (Chapter 3). The remainder of the workshop's first day was organized around three different encounters that one individual female patient had with the health care system over the course of a 15-year period and three life events. In the first (Chapter 4), she visits an obstetrician for preconception testing:

> In 2012, a 35-year-old Ashkenazi Jewish female smoker in good health is seen for a preconception visit. Under the current standard care model, targeted carrier status testing is offered. In terms of high effect sized variations that would be detected by traditional genetic testing, she is found to be a carrier for Tay-Sachs. In addition, if testing were extended in this scenario beyond what might be considered to be current standard of care, she would be found to harbor a prothrombin gene mutation, as well as variations in CYP2C9 and VKORC, indicating that she is likely to be highly sensitive to warfarin anticoagulation. She is also homozygous for ApoE4, but does not have familial hypercholesterolemia. She can be expected to have lower risk

variants and variants of unknown significance in accordance with expected population frequencies for the conditions under consideration.

In the second (Chapter 5), she develops a spontaneous deep vein thrombosis:

> The individual is seen at 40 years of age with progressive left lower extremity swelling and pain. Evaluation reveals an unprovoked deep vein thrombosis in her left lower extremity. She will be treated as an outpatient with low-molecular-weight heparin and warfarin. Targeted testing includes CYP2C9 and VKORC gene analysis.

In the last (Chapter 6), she develops a lung cancer:

> The individual is seen at age 50 with cough, dyspnea, and chest discomfort. Evaluation reveals a lung mass; bronchoscopy and biopsy reveal a non-small-cell lung cancer. Her tumor is found to have variations that allow the use of targeted therapy, and with treatment the patient goes into remission.

The three case scenarios were developed and presented to speakers to provide a guiding framework for discussions about the downstream and ancillary effects of providing genomic information in the clinical setting. The scenarios represent potential points where genetic information may currently provide value in clinical decision making and allow for a discussion of the potential sources of benefits and costs associated with three models of genomic data delivery:

- Targeted mutation detection using individual or panels of tests (current standard of care). This will include detection of variants of unknown significance.
- Whole genome sequencing with provision of data relevant only to the current clinical situation and a handful of high effect sized "actionable variants." This will include detection of variants of unknown significance.
- Whole genome sequencing with provision of data relevant to the clinical situation as well as other potentially significant secondary findings using the current best available data for interpretation. This will include lower effect sized variants, as well as variants of unknown significance.

Two separate panels reacted to each of these three scenarios. The first panel consisted of a clinician, a futurist, and a patient, who talked about how having genomic information could affect the choices, attitudes, and needs of stakeholders throughout the health care system. The second panel consisted of three economists who discussed the major economic issues surrounding the three scenarios.

On the second day of the workshop, the panelists from the first day reflected in a condensed form on their conclusions from the day before. Workshop participants also commented on the implications of issues raised during the workshop. These reflections and comments constitute the final chapter of this workshop summary.

MAJOR THEMES OF THE WORKSHOP

In his concluding remarks at the workshop, W. Gregory Feero, who at the time was a special adviser to the director of the National Human Genome Research Institute, offered his perspective on the major themes that emerged from the day and a half of discussion. Feero's summary of these themes is presented here as an introduction to the wide range of topics that arose in considering the economic consequences of genomic technologies. These ideas should not be seen as the conclusions of the workshop as a whole, but they do provide an overview of the topics summarized in the remainder of this volume.

The diversity of issues that comprise the economics of whole genome sequencing requires a spectrum of expertise and perspectives, Feero said. Some of these issues are solely economic, but others involve technology development; research needs; ethical, legal, and social issues and education; and health services. Each of these issues poses obstacles to the integration of genomics into clinical care and each needs to be well understood if the potential benefits of genomics are to be maximized.

Economic Issues

The economics of genomic sequencing vary by application and by setting, Feero said. A major question is therefore how to frame and analyze the economic issues. Values and costs can be measured in different ways, and these methods influence decisions about the use of technologies. In particular, improved methods are needed for assessing value, personal utility, and patient preferences.

A related complication is that public health, clinical care, and academic medicine have different economic assessment models. These models have to be aligned in a way that makes a difference to patients, said Feero. Also, particular models will be more or less useful in the currently evolving health care environment.

The infrastructure needs to be developed to measure outcomes related to economic factors along with standard health outcomes, not just for genomics but across the health care system. For example, better and quicker

INTRODUCTION AND OVERVIEW

approaches are needed for performing economic evaluations of genetic and genomic tests and the consequences of assaying particular genetic variants. Evaluating tests and variants one by one will be too daunting, said Feero. Sorting tests and variants into categories that can be assessed is one possible way of achieving this objective.

Economic analyses should be integrated into ongoing whole genome sequencing clinical studies, Feero said. It is being considered in some demonstration projects, but it could be part of all clinical studies. The economic incentives for test and evidence development under the current system of reimbursement versus a value-based pricing approach that incorporates the intellectual cost of interpretation need to be further explored.

If health care resources are flat or declining, and a potentially innovative technology is available, what or who will be replaced to allow for funding of genomic interventions? People will need to come to grips, said Feero, "with the fact that we should not be paying for very expensive, not particularly efficacious things in lieu of some things in genomics that actually are efficacious and not that expensive."

Technology Development

Sequencing will continue to get faster, cheaper, and more accurate, said Feero. At the same time, cheaper and faster technologies are needed for molecular characterization of samples beyond DNA.

Integrating genomic information into health information technologies and other infrastructures is constrained with current information technology systems. In academia, for example, many information technology departments have long lists of problems to solve and a finite budget, noted Feero, and these problems will compete against the incorporation of genomic results into databases.

Research Needs

Better methods are needed to determine which genetic variants should be acted upon in a clinical encounter. Behavioral research could determine if and how genomic information modifies the behavior of patients and health care providers, which is particularly important because this behavior will be a major driver of costs, said Feero. Also new methods are needed to increase participation in clinical trials, including participation of underrepresented subpopulations.

Epidemiological research is needed to evaluate risk assessments across platforms for various conditions, noted Feero. Epidemiologists also need to determine the relative contributions of environmental factors to health outcomes.

In general, resources need to be shifted toward translational research, said Feero, and this research needs to illuminate the economics of adapting new technologies.

Ethical, Legal, and Social Issues and Education

In the area of ethical, legal, and social issues, outcomes data on informed consent is a major need, cited Feero. What kind of informed consent is appropriate in the relationship between provider and patient?

In the area of education, Feero asked, can more efficient methods for patient and provider education be developed? Also, genomic scientists and clinicians need education about economic analyses applied to genomic tests.

Health Services

Health systems will need new methods and a stronger infrastructure, including informatics, to track and analyze the downstream consequences of providing sequence data, said Feero. For example, do codes exist that will follow what happens when genomic information is made available?

When should genomic sequencing be done during the lifespan of an individual, Feero asked. Possibilities range from having the complete sequence available at birth to conducting targeted sequencing at the time of diagnosis. If genomic results that are already available are more likely to be used than results that need to be obtained after the patient presents themselves, this raises the question of thresholds for the use and generation of evidence.

Knowledge gained from new technologies may not be applicable to all populations because not all populations are represented in research, noted Feero, which could heighten disparities in health care. Efforts should be invested in determining how new technologies could exacerbate or ameliorate existing disparities. However, it is important to remember that this issue is not specific to genomics.

Finally, asked Feero, in a world of stable or declining resources, do accountable care organizations provide a model for producing more efficient health care using genomic technologies?

The Need for a Systems Perspective

All these issues need to be considered from a systems perspective, said Feero. Most researchers, including economists, consider problems within a particular context and develop a carefully designed question, which produces an internally consistent and robust answer for that question. But any such problem is just part of a much larger overall picture. Particularly in health care, economic analyses encompass issues that range far beyond costs and benefits to complex issues of regulation, ethics, and equity, as the above themes demonstrate. Many different sources of information will need to be brought together efficiently to enable informed decision making and to determine how to move forward with integrating genomic medicine in a way that maximizes patient benefit while at the same time making the most economic sense.

2

Genomics, Population Health, and Technology

Important Points Made by the Speaker

- The incorporation of genomic sequencing into medicine will depend not just on the falling costs of genomic screening but also on the value that genomic sequencing provides.
- Genomic testing may have important implications for people with some diseases, such as familial disorders or progressive neurological diseases.
- For healthy people, genomic data are unlikely to have much effect on assessing the risk of common diseases.
- Nevertheless, genomic screening could be used to find the relatively rare individuals in a population who are at high risk of preventable disease, preemptively identify genetic variants that influence the effects of drugs, provide additional information for screening of newborns, and inform a variety of reproductive decisions.
- Genomic testing should be viewed as another available test and only used when and if the situation warrants.

Economics is not just about money, said James Evans, Bryson Distinguished Professor of Genetics and Medicine at the University of North Carolina at Chapel Hill, who provided one of the broad introductory talks that led off the workshop. Money is a proxy for the value people ascribe to something, and value is the fundamental concern of economics. The critical issue for new technologies, such as genomics, is therefore not just how much they cost but also how much value they produce.

THE VALUE OF GENOMIC DATA

The genome contains a tremendous amount of data, said Evans. Each individual differs at millions of genetic locations from the reference human genome. Some differences influence physical traits, such as eye color, while others influence medically important characteristics. Nonetheless, only rarely do polymorphisms greatly influence health. "It is important to keep that in mind," Evans said.

Evans divided genetic variants that affect health into two categories. In the first category are variants that occur frequently in the general population but have only a subtle impact on health. These variants raise the risk of a particular adverse health effect by only a modest amount, and geneticists do not yet know how best to aggregate such information to predict overall risk. These variants tend to have little utility in most clinical settings, said Evans.

In the second category are those variants that are found rarely in the population but that dramatically increase the risk of a health disorder. In these cases, the relatively "blunt tools" of modern medicine, such as bilateral mastectomies, annual colonoscopies, or drugs that can have substantial side effects, can be useful for preventing or treating disease on the basis of the knowledge gained from genomic information.

Because of its limited utility, genomic testing has not been widely adopted despite falling costs, said Evans. "I don't mean to say that this isn't marvelous technology, but we need to think about its utility to people before we rush to the conclusion that it is going to be, or should be, immediately embraced by everyone."

THE LONG-TERM AND MID-TERM PROMISES OF GENOMICS

Genetics will eventually shed light on the underpinnings of virtually every human disease, said Evans, because virtually every disease has a genetic component. In the long run, it undoubtedly will transform medical science.

But medical science is not the same thing as medical practice. Medical science is the indispensable foundation of medical practice, said Evans, but practice is far more complex than the underlying science. Theory alone is insufficient to guide practice, and the timeline for translation of science into medicine is long. Sickle cell anemia has been understood at the genetic level since 1949 (Neel, 1949; Pauling et al., 1949), yet treatment of patients has remained basically unchanged over this time. Medical practice is also far more expensive than medical science, and the stakes are far higher. "If you screw up, people literally suffer and die," Evans said.

Despite the gap between medical science and medical practice, Evans noted, a current application of whole genome sequencing is proving to be exceedingly valuable. For people who have a disorder with a genetic etiology, genomic diagnostics can provide tangible benefits by giving people information about their conditions that can be used to guide treatment or prevention measures. Evans cited genomic analysis of tumors as being a specific area where these benefits could be achieved in the near term. Moreover, even if no treatment for a condition is available, many people want a diagnosis. The information can end the "diagnostic odyssey" of patients going from physician to physician, trying to find out what is wrong with them, thereby reducing anxiety and saving resources. In some cases, this information can also inform reproductive decisions and direct preventive strategies for family members who may also be at risk.

Nevertheless, this application of whole genome sequencing will be useful in only a limited number of cases, said Evans, such as children with multiple malformations, familial disorders passed among multiple generations, progressive neurological disorders, and patients with unusual presentations, such as cancer at a young age. Most common diseases, such as diabetes or hypertension, have multiple causes, including factors such as diet, smoking, exercise, and the environment, and the contribution of any one genetic variant is small. This multifactorial etiology places an inherent ceiling on the utility of genetic testing for these disorders. "I don't think we are going to be able to get around that basic stumbling block and answer everything we want to know about, [for example], heart disease with genetic analysis," Evans said.

GENOMIC DATA IN HEALTHY PEOPLE

A different set of considerations surrounds the use of genomic tests in healthy people, said Evans. Healthy people have less to gain and more to lose from any medical intervention, including genomic tests.

Assessing the risk of common diseases through whole genome analysis of a healthy person has received the most attention, but this attention "is somewhat misplaced," Evans said. Currently, assessment of genetic risk alleles has "rather feeble predictive power" because the increased risks tend to be small. "From a clinical standpoint I don't know what to do with patients who are at a 1.3 relative risk for colon cancer," said Evans. "Am I going to hurt them by doing more intensive screening, or am I going to help them?"

In addition, few data suggest that knowledge of one's genomic status is effective in changing behavior. Moreover, even if it is, genomic data also could be a double-edged sword, said Evans, if individuals forgo healthy diets and exercise because of a perceived decreased risk of developing a disease.

"I know what almost everybody in this room is going to die of," said Evans. "We are going to die of heart disease or cancer. . . . We are all at high risk for these maladies regardless of our [genomically determined] risk. And many at decreased risk for heart disease will still die of heart disease. So we are all going to benefit from interventions that lower heart disease. We don't really need to target people. It doesn't do anyone much good to tweak our estimation of an individual's relative risk for common diseases which we are all at high absolute risk of developing anyway."

A possible application of genetic testing in healthy people is finding the relatively rare individuals in a population who are at high risk of preventable diseases—what another workshop participant called "newborn screening of adults." Risk assessment will always be most valuable when the identified risks are high. For example, about 0.2 percent of the population carries deleterious mutations that cause Lynch syndrome (Hampel et al., 2008), placing them at extraordinarily high risk for colorectal cancer, which is a preventable disorder. Today these individuals are identified only after numerous family members have developed cancer or died. Genomic testing could make it possible to do population screening for such disorders. Altogether, perhaps 1 percent of the population might harbor genetic variants that create dramatically increased risk, Evans estimated. "That is not small change. The number needed to treat for a lot of interventions, like statins for high cholesterol, is around that number for primary prevention," he said.

Preemptively identifying genetic variants that influence the effects of drugs in individuals is another promising application of genomic testing. Still, this application will probably be useful for a minority of drugs, Evans said. Even today, after years of development in pharmacogenomics, few loci have demonstrated unequivocal value in improving outcomes or reducing costs. Genetic testing to inform the use of abacavir (Mallal et al., 2008) is an exception to this generalization. But for other promising variants, data still are being collected regarding whether testing benefits patients. Furthermore, such testing may not require a genomic approach. Targeted genotyping at the point of care rather than advance knowledge may be preferable because pharmacogenomic information is only needed when a drug is prescribed. And retesting may be necessary for high-stakes decisions because test results can be wrong and because the tests themselves improve over time.

Genomic testing as an adjunct to newborn screening also holds considerable potential, said Evans. Genomic screening will not replace the current metabolic-based screening in the near term, because it remains closer to the phenotype of interest and has much greater specificity. For example, elevated phenylalanine has much more clinical utility than a variant of uncertain significance in the phenylalanine hydroxylase gene. But genomic screening could help resolve ambiguous biochemical results and detect a subset of treatable disorders that do not have good metabolic markers, such as storage diseases, deafness, and neonatal diabetes.

Finally, genomic tests can inform a variety of reproductive decisions, which is an area that Evans believes will "take off tremendously." Preconceptual carrier screening (see Chapter 4) is currently recommended for a few disorders, but these have been chosen essentially based on cost and mutation prevalence. Screening is conducted for cystic fibrosis or Tay-Sachs disease because it is affordable and because reliable testing is available, not because Tay-Sachs is any worse than, for example, Batten disease, said Evans. "That is not what couples really want to know. They want to know if [their] child is likely to have a really bad, untreatable disease." Genomic sequencing can help address these concerns by potentially being used to screen for all serious diseases.

Preconceptual carrier screening for serious diseases could have "a potentially profound and very welcome impact on family planning," said Evans. Some people will treat such information as highly actionable. Others will regard it as morally problematic. The formulation of policy in this area will be difficult, Evans warned.

CHALLENGES TO IMPLEMENTATION

Effectively harnessing genomic screening faces significant challenges. Because of the large number of bases in the complete haplotype genome, even an accuracy of 99.99 percent will produce 300,000 errors per patient, said Evans, though accuracy will gradually improve.

In addition, each person has about 4 million genetic variants, and our current understanding makes their interpretation difficult. Should information about all of them be gathered or stored? As genome sequencing becomes more accurate and cheaper, it may be more practical to do sequencing when the information is needed, Evans said.

Another significant challenge, said Evans, is that the genome is an unpredictable—and not necessarily friendly—place. For some people, whole genome sequencing will uncover things they were not looking for and might not want to know. Some people will discover that they are at high risk for untreatable and horrific conditions, such as fatal familial insomnia, Huntington's disease, or early-onset Alzheimer's disease. The potential for returning information when there is no medical action that can be taken is an important externality, Evans said, in deciding whether to do whole genome sequencing on everyone. Furthermore, different people will make these decisions differently, and these decisions are even more difficult when parents and children are involved.

Evans briefly described several social challenges to genomic screening. Genetic discrimination remains a concern. In the United States, the Genetic Information Nondiscrimination Act of 2008 protects against discrimination in medical insurance and in the workplace, but no such protections exist for long-term care insurance, disability insurance, or life insurance.

Widespread genetic testing poses the threat of allelism—that people will be defined by their genetic sequences and by the traits those sequences produce rather than by the qualities that truly matter in a person.

About 20 percent of the human genome has patent claims, which means that whole genome sequencing has the potential of being interpreted as violating multiple patents, said Evans.

Widespread testing would pose privacy issues because genomic information is digital and would be easy to distribute. Who will control and have access to this information?, Evans asked. People who volunteer for genetic tests can become upset, for example, if they learn that their genomic information is the property of a private company.

ANOTHER MEDICAL TEST

In the end, Evans concluded, whole genome sequencing is just another medical test. It is a highly complex test with great potential, but claims that everyone will undergo genome sequencing are based on high perceived utility and low cost, and for now only the low cost is being realized. "The old adage that an elephant for a nickel is only a bargain if you have a nickel and you need an elephant applies here. I am not sure most of us need that elephant. Even if free, perceived low cost is an illusion, because the misapplication of medical tests—and make no mistake, whole genome sequencing is a medical test—is very expensive," he said.

Genomic testing is likely to be applied as other medical tests are: when and if the situation warrants. Genomic analysis of a panel of variants could be useful in nondiagnostic settings. But Evans argued against burdening the health care system with a flood of extraneous information that cannot yet be interpreted and that may not be welcomed by many people. Ultimately, much more high-quality, outcome-based information on the uses of genomic tests is needed, he concluded.

3

The Intersection of Genomics and Health Economics

Important Points Made by the Speaker

- Health economics can provide a variety of tools and frameworks to help guide the implementation of genome sequencing in clinical practice.
- The majority of new medical interventions improve outcomes and increase cost.
- A cost-effectiveness framework for whole genome sequencing could consider the prevalence and penetrance of a genetic variant, the effects of that condition, the cost and accuracy of a test, the cost of an intervention, the outcomes of an intervention, and the severity of the disease.
- Greater understanding of patient-centered outcomes is needed to determine the value of genome sequencing.
- Patient and provider responses to genome sequencing will require new investments in services, new pathways of care, genetic counseling, decisions about what will and will not be covered by insurance, and provisions for dealing with incidental findings.
- Quicker approaches that incorporate qualitative assessments need to be devised for the economic evaluation of genetic information.

David Veenstra, professor in the Pharmaceutical Outcomes Research and Policy Program at the University of Washington in Seattle, began his overview of health economics by reiterating the point Evans made about health economics not really being about costs; rather, he said, it is about value and understanding utility. Through a consideration of value, health economics can help people clarify the assumptions that are being made, consider uncertainties, and evaluate trade-offs. Thus, health economic evaluations are primarily used to inform decision making.

A basic tenet of economics is that people make decisions to improve their well-being. For most commodities, price is a measure of perceived improvements in well-being, or value, and people make decisions on the basis of value. These principles, however, often do not apply in health care, Veenstra observed. The individual receiving the health care is generally not the person who makes the decision about what health care will be received. The person receiving the health care typically does not have a good idea of the potential benefits and harms of a decision. And patients generally do not pay out of pocket for the services they receive.

Health care economics tries to gain a better understanding of the value of one health care intervention compared to an alternative approach, taking into consideration all the impacts across patients, providers, and the health care system. This value can be measured in terms of price, improvements in quality of life, a longer life expectancy, resources saved, health state, and so on. The key components of the evaluation, Veenstra said, are that all relevant factors are included in the analysis and that the same approach is applied to all decisions that are being assessed.

ECONOMIC EVALUATION TOOLS

Economists use several different tools to carry out economic evaluations of health care interventions, including cost-minimization analysis, cost-benefit analysis, cost-effectiveness analysis (CEA), and cost-utility analysis (CUA). All these approaches consider the cost of the intervention as well as downstream costs, but they differ in how they measure the outcome or utility of an intervention (see Table 3-1).

Cost-minimization assumes that the outcome of two different interventions is the same. The goal in this case is to reduce costs, and thus the cheapest intervention with the same effect on outcome can be determined. Still, outcomes tend to be different to some degree, said Veenstra, so this approach cannot be commonly used.

Cost-benefit analyses consider everything in terms of costs. But this can be difficult to do in health care, Veenstra said, because people tend to resist putting a monetary value on health, and thus it is extremely hard to measure accurately.

TABLE 3-1 Types of Economic Evaluations in Health Care

Study Design	Costs Measured?	Outcomes Measured?	Strengths	Weaknesses
Cost-minimization	Yes	Not necessary	Easy to perform	Useful only if outcomes are the same for both interventions
Cost-benefit	Yes	Yes, in monetary terms	Good theoretical foundation; can be used within health care and across sectors of the economy	Less commonly accepted by health care decision makers; evaluation of benefits methodologically challenging
Cost-effectiveness	Yes	Yes, in clinical terms (events, life years)	Relevant for clinicians; easily understandable	Cannot compare interventions across disease areas when using disease-specific end points
Cost-utility	Yes	Yes, in quality-adjusted life years	Incorporates quality of life; comparable across disease areas and interventions; standard	Requires evaluation of patient preferences; can be difficult to interpret

SOURCE: David Veenstra, IOM workshop presentation, July 17–18, 2012.

The two more commonly used approaches in the field, said Veenstra, are CEA and CUA. CEA is a quantitative framework for evaluating the complex and often conflicting factors involved in the evaluation of health care technologies. CEA seeks to determine whether an intervention used to prevent, diagnose, or treat an illness improves clinical outcomes enough to justify the additional dollars spent compared with alternative uses of the same money. CEA is not a method to show which interventions reduce cost. Rather, it aims to inform which interventions provide the greatest value for the amount of money that is spent. Also, CEA is not a method that removes individual or group responsibility for making clinical and financial decisions. Rather, it provides information that is incorporated into larger decisions involving additional considerations, such as issues of equity.

CEAs measure outcomes in terms of clinical events such as cost per heart attack avoided or cost per life year saved. This approach works well, said Veenstra, and people are reasonably accepting of it. CEAs, however, do not easily allow for cross-intervention comparisons, for example, whether to spend $50,000 to prevent a heart attack or spend $50,000 to prevent breast cancer. Answering this question would require further considerations of how long a person would have lived, what that person's quality of life would have been with a given intervention, as well as other downstream costs.

The gold standard in the field has become CUA because of this limitation of CEA analysis, said Veenstra. CUA typically measures outcomes through a metric called a quality-adjusted life year (QALY) and allows for comparisons across interventions. For example, if $50,000 spent to prevent a heart attack produces 10 QALYs, and $50,000 spent to prevent a breast cancer produces 20 QALYs, a decision can be informed by that information. "That is what we produce in health care," said Veenstra. "We don't make cars; we don't make phones. We increase people's length of life, and we improve their quality of life—at least that is our goal. And the QALY captures those."

INCREMENTAL COST-EFFECTIVENESS RATIOS

Another standard measure in health economics is the incremental cost-effectiveness ratio, which is defined as the difference in cost between two interventions divided by the difference in their effectiveness. This metric can fall into four different quadrants on what is called a cost-effectiveness plane (see Figure 3-1). The best result is when outcomes improve and costs go down. The worst is when outcomes become worse and costs increase. Most interventions in health care result in higher costs with improved outcomes, Veenstra said, which makes CUAs useful for comparing these interventions. For example, the cost per life year saved may be $10,000 for one intervention and $200,000 for another intervention. In this case, money may be more effectively spent on the first intervention.

In the United States, however, there is not a clear threshold on how much money society is willing to spend to save a life for 1 year, said Veenstra. "You might hear people [say] $50,000 per QALY. In reality, it is probably closer to $100,000 or more in this country." Nevertheless, this approach provides a way to determine whether an intervention is reasonable.

GENOME SEQUENCING

Whether genome sequencing is cost-effective depends on the outcome that is being measured, Veenstra said. These outcomes could be measured in

THE INTERSECTION OF GENOMICS AND HEALTH ECONOMICS 19

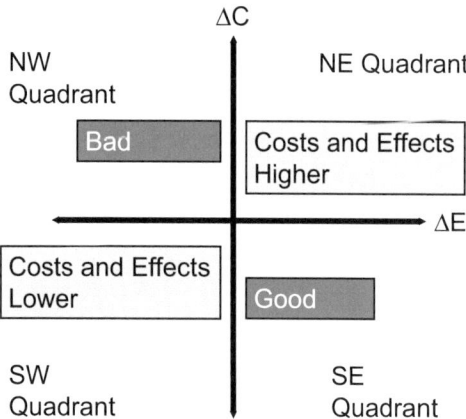

FIGURE 3-1 The change in costs and change in effectiveness compared with current practice divides the results of cost-effectiveness analyses into four quadrants.
NOTE: ΔC, change in cost; ΔE, change in effectiveness.
SOURCE: David Veenstra, IOM workshop presentation, July 17–18, 2012.

terms of base pairs sequenced per dollar, the number of clinically meaningful genetic variants identified, diagnoses received, clinical actions taken, or patient outcomes. The other important factor is the comparator. Is genome sequencing being compared to nothing, to observing the patient in the clinic, or to a targeted sequencing approach?

Flowers and Veenstra (2004) developed a framework for factors that could influence cost-effectiveness in pharmacogenomic testing (see Table 3-2). Important factors include the prevalence and penetrance of the genetic variant, the cost and accuracy of the test, the prevalence of the disease and the outcomes if left untreated, and the effectiveness and cost of treatments. A similar framework could be constructed for whole genome sequencing to examine benefits and harms, according to Veenstra. That framework would consider the prevalence of the variant of interest, the penetrance of the condition, the cost of a test, the cost and outcomes of an intervention, and the severity of the disease. A major complication is that a typical economic evaluation of a single test or single genetic variant can take a year. "We don't have time for that," Veenstra said. "We need to have quicker approaches that use more of a qualitative assessment." Yet, if enough examples of this type of analysis can be completed, he added, "we can get a sense of where good value may be provided."

TABLE 3-2 Factors That Influence the Cost-Effectiveness of Genomic Testing Strategies

	Factors to Assess	Features That Favor Cost-Effectiveness
Gene	Prevalence	• Variant allele is relatively common
	Penetrance	• Gene penetrance is high
Test	Sensitivity, specificity, cost	• High specificity and sensitivity • A rapid and relatively inexpensive assay is available
Disease	Prevalence	• High disease prevalence in the population
	Outcomes and economic impacts	• High untreated mortality • Significant impact on quality of life • High costs of disease management using conventional methods
Treatment	Outcomes and economic impacts	• Reduction in adverse effects that significantly impact quality of life or survival • Significant improvement in quality of life or survival due to differential treatment effects • Monitoring of drug response is currently not practiced or difficult • No or limited incremental cost of treatment with pharmacogenomic strategy

SOURCE: David Veenstra, IOM workshop presentation, July 17–18, 2012.

COMPARATIVE-EFFECTIVENESS RESEARCH

Comparative-effectiveness research, which is an amalgamation of previous approaches in technology assessment and health economics, also relates to the issue of how people use information from genomic tests. According to Veenstra, comparative-effectiveness research includes all of the following components:

- Stakeholder-informed prioritization and design of studies
- Direct, head-to-head comparisons
- A broad range of beneficiaries, including patients, clinicians, purchasers, and policy makers

- Study populations representative of clinical practice
- A focus on patient-centered decision making

The emphasis on patient-centered outcomes in comparative-effectiveness research is especially relevant to whole genome sequencing, Veenstra said. The new Patient-Centered Outcomes Research Institute plans to fund studies that investigate the key determinants of outcomes patients experience following treatment decisions, with a special emphasis on studies considering that results may differ among patient groups on the basis of patient characteristics. The center also will be supporting studies that compare the use of prognostication and risk stratification tools with usual clinical approaches. "These are the types of things that we are going to need to get a better understanding of the value that whole genome sequencing brings," Veenstra noted.

THREE CHALLENGES

Veenstra cited three challenges at the intersection of genomics and health economics. The first is whether decreasing sequencing costs can reduce the incentives for test development. If whole genome tests cost less than $1,000, individual genetic tests may no longer make sense. At that point, does the provision of clinical interpretation become the value proposition, with each condition a patient presents with eliciting a new review of relevant whole genome sequence data? And if so, are reimbursement systems designed to reward this? As Ramsey et al. (2006) noted, value-based payment policies would provide greater incentives for continued innovation in test development, Veenstra said.

The second challenge involves the development of policies to incorporate personal utility into test assessments. Diagnoses do not necessarily increase life expectancy, but individuals may value this information. What is the value to an individual of knowing something, asked Veenstra. Standard approaches in health economics are limited in that they do not sufficiently capture patient-centered outcomes. New techniques and tools, such as conjoint analysis and discrete choice experiments, have been developed to get a better sense of the value of knowledge to patients (Basu and Meltzer, 2007; Grosse et al., 2008; Regier et al., 2009). These tools need to be further developed so that consideration can be given regarding whether to incorporate such measures into guidelines or into reimbursement policies.

A related policy issue involves evidence thresholds. Does a lower cost of obtaining information lead to a lower evidence threshold for using that information?, Veenstra asked. For example, he said, genetic testing for warfarin dosing is rarely done today even after 10 years of evidence development, but if a test result were already available, clinicians would likely take the result into account in choosing an initial dose. He noted, "There is this issue that if it takes time and money to get the information, it is not worth it. But if you have it sitting there in front of you, you are going to go ahead and use it." In an environment of limited evidence, an economic tool called value-of-information analysis may provide a framework for "evidence-based" decision making. This tool gives a sense of whether having information today is worthwhile as compared to developing the evidence base further.

The third and final challenge Veenstra described is the impact of whole genome sequencing on the health care system. We do not know how patients and providers will respond to the information available from genomic tests, specifically as it impacts their decisions about receiving medical care. Any response may necessitate further investment in services, the development of new pathways of care, genetic counseling, decisions about what will and will not be covered by insurance, and provisions for dealing with incidental findings.

4

Preconception Care and Sequencing

> **Important Points Made by the Individual Speakers**
>
> - Lack of insurance coverage greatly affects the access to and use of preconception genetic testing.
> - The increasing number of genetic tests is producing significant educational challenges for providers and patients, which is exacerbated by insufficient numbers of genetic counselors.
> - Genetic variants of uncertain significance can present major challenges in the preconception and prenatal period and produce extremely difficult counseling scenarios.
> - Secondary findings from a genomic test must be dealt with on a case-by-case basis.
> - Genomic data can provide direct information about a patient, but it also can have relevance to family members as well.
> - A framework should be developed to put value on the information being provided by genetic tests.
> - New or revised tools for more effectively conveying information to patients will need to be developed in order to accommodate the significant amount of information from genomics-based testing.
> - Using genomics as a tool for avoiding unnecessary procedures or treatments has significant potential to save costs for the health care system.

In the first scenario discussed at the workshop, a woman who is contemplating pregnancy seeks counseling:

> In 2012, a 35-year-old Ashkenazi Jewish female smoker in good health is seen for a preconception visit. Under the current standard care model, targeted carrier status testing is offered. In terms of high effect size variations that would be detected by traditional genetic testing, she is found to be a carrier for Tay-Sachs. In addition, if testing were extended in this scenario beyond what might be considered to be current standard of care, she would be found to harbor a prothrombin gene mutation, as well as variations in CYP2C9 and VKORC, indicating that she is likely to be highly sensitive to warfarin anticoagulation. She is also homozygous for ApoE4, but does not have familial hypercholesterolemia. She can be expected to have lower risk variants and variants of unknown significance in accordance with expected population frequencies for the conditions under consideration.

A CLINICIAN'S PERSPECTIVE

Siobhan Dolan, associate professor of clinical obstetrics and gynecology and women's health at the Albert Einstein College of Medicine, noted that 50 percent of pregnancies are unplanned, so most women do not come in for this type of clinical assessment prior to conception. She also noted that when a woman is not pregnant, many insurers will not cover genomic screening. "For many of our patients, they don't have access if it is not covered," she said. Even when a patient has insurance coverage, clinicians have to spend time checking to be sure which genomic tests are included in that coverage. "That is a rate-limiting component of access for many women in this country," Dolan added.

Targeted Mutation Testing

There is an immediate decision point for the patient and provider on what screening to perform for targeted testing, said Dolan. The American College of Obstetricians and Gynecologists (ACOG) recommends screening for nine conditions that are more common among Ashkenazi Jewish populations. Certain philanthropic programs, however, are currently offering screening for 19 conditions, including diseases in which there is a higher risk in this population, such as cystic fibrosis, as well as diseases in which there is not an elevated risk but for which screening is generally recommended, such as spinal muscular atrophy. The number of tests offered has a tendency to rise over time, said Dolan, which presents educational challenges as well for both providers and patients in understanding the appropriate use of new tests and being able to interpret and act upon results.

With a personal or family history of autism spectrum disorder or intellectual disability, testing for Fragile X syndrome would be considered. Though this patient did not seem to have either of these conditions, some researchers call for offering this screening to everyone, Dolan said, but guidelines today require a history before doing so. Likewise, for spinal muscular atrophy, ACOG recommends offering screening only to women with a family history, whereas the American College of Medical Genetics and Genomics recommends it for all women. "It is difficult for clinicians to act in that setting," said Dolan. In addition, Dolan would discuss maternal age and the risk for aneuploidy.

One in four Ashkenazi Jewish individuals will be found to be a carrier of at least one condition through genetic screening. This high incidence rate points toward the value of screening in this population. Dolan cautioned, however, that DNA screening may not be appropriate for other high-risk groups because current mutation testing may not be effective outside of Ashkenazi Jewish populations. Dolan recommended that enzyme testing and not DNA screening be offered instead for these individuals. She also suggested that Jewish groups should get enzyme testing in addition to genomic screening, which would present a challenge for whole genome testing. Many genomic tests have enzyme tests used as an adjunct to screening, and if large numbers of conditions were being tested, this adjunct testing could become onerous.

If the woman is identified as a carrier, the partner needs to be tested to have any impact on the pregnancy and the health outcome. But for various reasons, partners often cannot be tested. For example, they might be out of the country for work, overseas in the military, incarcerated, or uninsured and unable to pay for testing. "When we don't have partners, we certainly increase the maternal anxiety in the pregnancy. We certainly spend lots of money. We don't necessarily improve any health outcomes. It is a side effect to keep in mind," Dolan said.

Genetic variants of unknown significance can present major challenges in the preconception and prenatal period. Women need to make very difficult reproductive decisions, which can produce extremely difficult counseling scenarios. Dolan noted, "We traumatize many, many women. These pregnancies become incredibly stressful, whether the outcome is good or bad."

Dolan said that she would focus on informing the patient of the availability of testing options and supporting autonomy in her decision making. She also would offer partner testing for Tay-Sachs. If the partner is Ashkenazi Jewish and is negative for Tay-Sachs mutations, he has a residual risk of 1 in 560 of being a carrier, she said. He should also be offered the enzyme assay, because 11 percent of Ashkenazi Jewish carriers will be missed if enzyme testing is not done, which will bring his residual risk of being a carrier to 1 in 1,451 if both tests are negative. The residual risk to the fetus then needs to be reported to the couple: the risk is 1 in 5,800 if the partner has full testing, and it is 1 in 2,240 if he has mutation testing alone.

If both partners are carriers, further counseling is needed. Preimplantation genetic diagnosis and in vitro fertilization is an option for some, but it may be too expensive for others. Additionally, some couples will choose to have an affected pregnancy, which is "an interesting challenge to this whole scenario," Dolan said. Is the goal of testing to produce informed choice or to decrease the rate of affected pregnancies? This is a deep ethical question at the core of the scenario, said Dolan.

The biggest impact a clinician could have on this patient is to help her quit smoking, Dolan observed. A meta-analysis of 20 prospective studies on preterm delivery comparing any maternal smoking versus no maternal smoking found an odds ratio of 1.27, signaling an increased risk of preterm birth with smoking (Shah and Bracken, 2000). Furthermore, prematurity creates huge health care costs (Russell et al., 2007), and it is a very common outcome, representing more than 12 percent of births (Martin et al., 2012).

Targeted Results Plus Actionable Variants

The second model for the provision of information envisions whole genome sequencing with the return of data relevant only to the current clinical situation and a handful of "actionable variants." In this case, Dolan suggested that an expanded panel of testing for carrier conditions could be offered, such as the Counsyl Universal Genetic Test, which includes more than 100 conditions (Srinivasan et al., 2010). Testing generally starts with the mother, and the father is then tested for either the full panel or conditions for which the mother was a carrier. These conditions have varying disease prevalence, and the sensitivity and specificity of the testing for each condition vary widely, so each disease has to be considered separately if a result is positive.

This particular patient is reported to harbor a prothrombin gene mutation, which increases her risk of thrombosis, the development of blood clots. The woman's risk for venous thromboembolism per pregnancy with no history is less than 0.5 percent. If she has had a previous venothromboembolic event, however, her risk would go up to 10 percent; prothrombin gene heterozygotes account for 17 percent of all venous thromboembolisms (Lockwood et al., 2011). On the basis of this information, Dolan would also want to test her Factor V Leiden mutation status because this mutation has also been associated with increased risk of thrombosis and "60 percent of venous thrombosis cases in pregnant women" (Grody et al., 2001). Still, a major question is whether genetic information can signal risk before a sentinel event, but no data are yet available to make that determination.

"With no prior events entering pregnancy, there is really no clear guidance [on] how to treat her other than to watch," Dolan said.

She would also ask the woman whether she wanted to know about her possible BRCA1 and 2 mutations. Most women, however, do not want to talk about breast cancer while they are thinking about pregnancy. Dolan said, "When you are excited about your new pregnancy, while it is true that your mother and your sister had breast cancer, you may not be receptive to that information at that time."

Dolan would also suggest possibly examining CYP1A1 and GSTT1 status because one study of 700 women demonstrated that specific genotypes modified the association between maternal cigarette smoking and infant birthweight, suggesting an interaction between metabolic genes and cigarette smoking (Wang et al., 2002). Eleven women with this relatively rare genotype delivered on average 5 weeks earlier. Dolan asked, If the woman had this genotype, could this evidence provide impetus to help her quit smoking? "This is research data, but I do think it could help us target who is at risk and what we could offer them," she said.

Many of these conditions are rare, so Dolan said that if she had a patient with a particular test result, she would seek out an expert on the natural history of the condition. But given that people all over the country need counseling, rare conditions are an additional challenge for the limited number of counselors available. A related question is how to reimburse genetic counselors and other staff for the huge amount of time that will be spent counseling patients about tests and the results of those tests.

A challenge with this expanded panel is that Counsyl is currently not accepting New York State Medicaid, so it is not affordable for some women, even though the $350 charge is a cost-effective way of testing for all these conditions. The question then becomes whether such testing will widen disparities if certain segments of the population will not have access.

As specific follow-up to the expanded genetic information, Dolan said that she would do a hematology workup and continue to emphasize smoking cessation. She would also consider anticoagulation therapy early in pregnancy, although no clear guidelines exist for the management of these patients, particularly in the absence of any prior event.

The Whole Genome Sequence—What Are We Paying For?

In the third and final model, whole genome sequencing is conducted, and data relevant to the current clinical situation as well as other potentially significant secondary findings are made available to the patient with the best current data for interpretation. Dolan said that as a provider, she would not be excited about having a lot more information. The information will include variants with lower effect sizes and of unknown significance.

Will the woman want to know that she is an ApoE4 homozygote? What if the emotions associated with learning that information affect her decisions about having children? Did her family history reveal any information about the penetrance of that variant? Did her family history suggest any other potential risks? For example, diabetes is a huge public health challenge in the Bronx, which leads counseling strategies toward exercise and diet without genetics playing a substantial role.

The economics of genetic testing can be very difficult, Dolan concluded, because "we don't really know what we are paying for." Informed decision making is a laudable goal, but testing and counseling are expensive and will not necessarily lead to fewer affected infants. Nevertheless, the amount of disease that can be prevented is tremendous, as is the excitement surrounding genomics.

A FUTURIST'S PERSPECTIVE

Arthur Beaudet, Henry and Emma Meyer Professor and chair of the Department of Molecular and Human Genetics, Baylor College of Medicine, discussed the case from the perspective of a futurist, looking at the kinds of capabilities and information that might be available 20 years from now. At that point, whole genome sequencing conducted in the first trimester using noninvasive techniques is likely to be common, he said. For the woman in the scenario, such testing would identify risk to her offspring caused by inherited conditions. It also would identify genetic risks related to new mutations, such as trisomies, point mutations, and deletions or duplications.

Beaudet divided the effects of genetic mutations into two categories. In the first category are debilitating conditions where individuals cannot live fully independently. Individuals affected in this way typically cannot advocate for themselves.

The second category includes conditions with milder severity. For these disorders, preimplantation genetic diagnosis becomes more of an option, said Beaudet. Examples might include breast cancer mutations, hereditary deafness, and achondroplasia.

From an economic perspective, whole genome sequencing will be more expensive than targeted testing, at least initially, but could be cost-effective if very expensive conditions are avoided, Beaudet said. If information provided to a patient or family from a whole genome sequence is limited, that restriction will be done not by designing a different test but by limiting the information to be shared. Also, from a multigenerational perspective, whole genome sequencing is far more cost-effective. If everyone has a whole genome sequence done at birth, it can be used throughout life, and the relevant information can be applied to other family members.

Beaudet said that it will be important to identify the causative mutations for all individuals with serious Mendelian disorders. This information will allow for a better understanding of the type of variation found for each disease and the clinical utility of identifying various mutations.

De novo mutations that are not present prior to conception, such as trisomies, genetic deletions or duplications, and point mutations, can only be detected through intrapartum testing. Such testing, however, could also produce secondary findings that would pose a challenge to patients and providers. For example, Beaudet said, "we are doing now quite a large amount of prenatal diagnosis using copy number arrays where we encounter copy number variants of unclear significance."

These secondary findings must be dealt with on a case-by-case basis, he said. He also stated that, in his opinion, more information is almost always better. He added, however, that "this is a personal opinion and not one I recommend for everybody." But a physician giving a physical does not avoid listening to a patient's heart because of the possibility of hearing a heart murmur. "We have the information that comes with the society and the technology that we currently live in," he said.

The behavior of providers must be regulated through provisions such as the 2008 Genetic Information Nondiscrimination Act to prevent abuses. But most patients can be counseled through informed decision making, said Beaudet, even with findings of uncertain significance. Interpretation and annotation, though expensive today, could drop in price as informatics develop. Nevertheless, delivering information to patients will almost certainly involve considerable time and resources.

Beaudet concluded by pointing out that the pediatric community already has considerable interest in whole genome sequencing. The Baylor College of Medicine began offering whole genome sequencing in November 2011, and after several months it received between 10 and 20 samples per month. The most recent month saw 69 samples. Most were from children, but a few were from adults who were looking for an underlying genetic cause for a disease. In about 30 percent of the samples, testing is revealing a disease-causing mutation. The use of whole genome sequencing "seems to be on the rise," he said.

A PATIENT'S PERSPECTIVE

The woman in the scenario may not know what to expect, said Michelle Gilats, a genetic counselor at the Chicago Center for Jewish Genetics. Though she is likely to have at least heard of Tay-Sachs disease, the patient is unlikely to know much, if anything, about the other conditions for which she is being tested. She is likely to expect that the testing will tell her whether her child will be at risk for certain conditions about which she may have to make different reproductive decisions. But she may not know what her options are if she is found to be a carrier, and she may not be aware that options exist. Also, she may not be expecting to receive genetic information about her own health, especially about conditions for which the implications are unclear and the significance unknown. Because of this lack of knowledge, pretest counseling is imperative, said Gilats.

Preconception screening to date has been determined mainly by ethnicity. But not everyone knows his or her ethnic background, and many people have mixed ancestry. This situation creates an advantage for larger screening panels because they are more universal in scope and reduce the need to rely solely on patients' self-assessment of their ethnicity.

The Center for Jewish Genetics uses the Counsyl panel for testing but gives people the option of doing a more limited Ashkenazi panel for 18 conditions. Most people choose the larger panel because it provides more information at the same cost. The downside is that a large panel can provide too much information. For example, it can produce results for conditions that do not have clear-cut responses, such as hereditary hearing loss, or conditions or traits that are not lethal or may not be very life altering. "Yet, because people have the information, they [can] feel they need to act on it," Gilats noted.

Gilats said that she used to work in prenatal genetics and often encountered patients who had maternal serum screening in pregnancy without being fully informed of the implications of possible results. When she would explain that the results indicated an increased likelihood of Down syndrome, patients could become angry because they had no intention of changing their pregnancy plans and would have refused the test if they had known what it might tell them.

Most patients have a poor understanding of genetics and the residual risk, said Gilats. Even in the well-educated population with which she works, the concept of residual risk for recessive disorders after carrier screening is often misunderstood. For instance, she recently told a patient that she was a carrier for a condition and that her husband was not, yet the patient was still not sure whether she should be concerned. Another patient, a doctor, was confused after being told that she and her husband still had a risk of having a child with cystic fibrosis even though she screened negative. And because many of these conversations occur in a clinical setting after a pregnancy occurs, rather than before conception, decisions made on the basis of this information can be even more difficult.

This lack of understanding will only be exacerbated with whole genome screening, said Gilats. It will not be possible to educate people about all the different disorders and results that are possible. Rather, explanations will have to be broad, outlining categories and examples of results. Having whole genome information for the woman in this scenario, for instance, would entail determining whether she would want to know her risk of developing certain cancers or currently untreatable diseases such as Alzheimer's. She would also need to be informed that the results could impact family members as well. Her options would need to be discussed prior to testing and a plan put in place for how, which, and when results would be delivered.

A participant noted that genomic data inevitably raise specific issues involving families. Does a patient want other family members to know about a genetic condition? What responsibility does a physician have to relay information to other members of a family? How will payers respond to various uses of this information?

Different testing and delivery models will lead patients to different actions. Having more information causes patients to ask more questions and spend more time with their providers, discussing their options and recommendations, said Gilats. "The hope is that information can be acted upon, such as with lifestyle modifications or medical intervention. But this won't always be the case," she said. Most people need little evidence to be concerned about a specific mutation, whereas a great deal of information is needed to reassure them that their concern is unwarranted.

Whether a result constitutes enough information to cause patients to change behavior remains to be determined, said Gilats, and depends on the specific condition. With conditions where the effects are not known or where a prevention or treatment cannot be recommended, the action a patient would take is even less clear.

In the second and third delivery models, the patient would receive results beyond autosomal recessive disorders. Depending on her specific results, she may need to see a specialist and follow up with extra surveillance or management. What are the costs of this follow-up? Would it be covered by insurance? Even if it is, the copayments alone may be cost prohibitive for some patients, Gilats said.

Costs may also be incurred by family members, either because they also are at risk or because they become a means to further assess a variant of unknown significance. Another potential cost is unnecessary screenings for surveillance purposes, said Gilats. If the patient has a mutation that increases her susceptibility to a common disease such as cardiovascular disease, but there is no family history, are extra screenings warranted?

Interpretations may and likely will change for some variants, thereby changing a patient's risk over time. Changes in the assessment need to be conveyed to patients, which will require reinterpreting and recontacting patients after the initial results are delivered, said Gilats.

Whole genome sequencing has the potential to deliver great benefits to patients, but the results need to have meaning, Gilats concluded, and the patients need to want the information that the testing can provide. "With great power comes great responsibility," she said.

ECONOMIC PERSPECTIVES

Scott Grosse, research economist and associate director for health services research and evaluation in the Division of Blood Disorders, National Center on Birth Defects and Developmental Disabilities, Centers for Disease Control and Prevention, pointed out that measures of cost-effectiveness and value to patients will not necessarily coincide. They will in a case such as screening for Tay-Sachs disease, where doing so in Ashkenazi Jewish populations is cost-effective and there is high value for the patient, but they may not with thrombophilia testing, where decisions on whether to prescribe anticoagulation medications are not necessarily driven by a risk-benefit balance. Grosse noted that more predictive information regarding thrombosis can be gained by looking at blood type, with Type A or Type B individuals being at two to four times higher risk of developing blood clots, than testing for rare variants such as Factor V Leiden or prothrombin (Dentali et al., 2012; Jick et al., 1969; Medalie et al., 1971; Sode et al., 2013). Nonetheless, blood type data are not factored into decisions for managing patients. The value of information, said Grosse, depends entirely on how and if it is used.

Scott Ramsey, full member in the Cancer Prevention Program, Division of Public Health Science, Fred Hutchinson Cancer Research Center, discussed how the tests mentioned in the scenario might be considered, reimbursed, adopted, and used within the current framework by which most health plans evaluate new technologies. Health plans do review evidence for genomic tests, but the decision to review is usually based on the cost of that test rather than on its purported benefits. For example, BRCA1 and 2 testing is carefully tracked by health plans because it is very expensive. But testing for individual variants such as CYP2C9 "falls under the radar" for health insurance plans because it is relatively inexpensive. If whole genome screening were to fall to a very low price, it could be below the level where health plans have the tools to identify it, though additional and possibly sizable costs will be associated with interpreting and annotating the information because these costs are not predicted to fall at the same rate as sequencing. An increase in use would also raise questions about data storage, noted one participant, particularly about who would house this information and pay for that service.

If whole genome sequencing is recognized by health plans, a major question is how it will be reimbursed. Some tests have codes that are used to identify and pay for them, but most do not, and these are often crosswalked against other tests with existing codes to make reimbursement decisions.

If genomic testing does fall under the radar, health insurance plans will see the consequences of that testing in the form of subsequent tests or procedures. "That is going to be the hardest thing that health plans are going to have to deal with," said Ramsey. "They aren't going to be able to pick up the individual test happening, but they are going to see all the downstream impacts in terms of health system use."

Finally, Ramsey observed that as the costs of sequencing tests continue to fall, multiple companies will be competing with each other to sell this service. One way they will distinguish themselves is through the number of variants that they report, which will create an incentive to create and generate increasing amounts of genomic data for each person. Paul Billings, chief medical officer for Life Technologies, added that in addition to services provided, companies will compete on experience.

Innovation in genomics has been unrelenting, said Billings, driven by unmet needs and current opportunities in the market. Much of that innovation is going on in industry, which has a responsibility to figure out ways to profit from its innovation. James Evans cautioned against allowing market drivers to determine policies for implementing genomic medicine, however.

Veenstra observed that comparative studies can have great value—for example, whole genome sequencing versus standard of care. Understanding the differential impact of using various approaches can provide fundamental information on their value.

The economists also discussed consumer preferences for more versus less information in genetic testing. Some portion of the population will want as much information as possible, while others will resist even information that has high value. To date, demand has existed for tests offering more information, but continued demand will depend on the cost of the testing, who is paying for it, and the consequences of testing.

Ramsey pointed to a coming crisis caused by conflicts between the need for genetic counseling and the resources available for those services. "As these tests provide more and more information, something is going to give. You can't provide more and more counseling given the limited reimbursement available," he said. The question then becomes where the people being tested will turn for additional information. Will companies provide that service? Will people use the Internet? Will genetic counselors refer patients to other sources of information? Will the information that patients receive be accurate? And will they be able to make sense of the information? "It really raises a lot of problems," Ramsey noted. He argued that better tools are needed to convey information to patients in ways that maximize their welfare.

As part of this conversation, Grosse pointed to some of the problems with using QALYs as a measure of health states. They do not necessarily do a good job of measuring people's willingness to make trade-offs. Also, they do a poor job of measuring acute end points instead of chronic end points—for example, people may pay a lot to avoid a 3-day food-borne illness, but that health state has virtually no effect on QALYs. And they do not measure many of the things people care about, such as a slight depression of IQ. Grosse asked, "Does that mean that preventing mild cognitive loss has no value to society? Of course not. It is just that the QALY is not designed to capture that type of end point. That's why we need a fuller set of tools."

One of the unheralded potential benefits of genetics will be to indicate when something does not need to be done, Ramsey said. For example, many patients who are diagnosed with low-grade, local-stage prostate cancer are treated aggressively despite the fact that 5-year survival for that cohort is very high. "Why are we doing that? We are doing that because men are worried about it, and there is an incentive for urologists to do that. If we could come up with genomic tests that told us with a high degree of certainty that that person was not going to go on and develop advanced prostate cancer, that would save the system billions and billions of dollars," Ramsey said.

Innovators could also help remove waste and cost from the system by identifying areas in which more precision at the same or reduced price could be found, added Billings. This could be achieved by replacing human variance with quantitative measures.

DISCUSSION

One participant observed that genomic medicine is evolving within the context of a changing health care delivery system. It would behoove the system and the underlying economics to begin to change to incorporate genomic medicine because eventually it will be part of standard medical practice, whether 10 years or 50 years from now.

Ned Calonge, president and chief executive officer of The Colorado Trust, pointed out that continued expansions of coverage to include items that are cost-effective will produce some improvements in health. Nevertheless, it will also increase the total cost of health care, resulting in "a system we can't afford." Thus, the decision to pay for something can have a negative impact by potentially reducing access to care for everyone, especially disadvantaged populations.

Grosse pointed to a disconnect between economics and reimbursement decisions. Many current health care practices are not cost-effective or even based on much evidence, he said. To control health care costs, it would be better "to stop doing things for which there is limited evidence of effectiveness rather than trying to prevent the adoption of new technologies where there is good evidence."

Calonge noted that newborn screening only looks at a subset of the genetic variants that can be detected using current testing technologies. It may be cost-effective to add variants to the screening panel, but this would add costs for additional testing and interventions as well as bring up issues regarding uncertainty about the effects of the variant being detected. The ultimate problem is that "I know what I am spending, but I don't know what I am buying," Calonge said. Genomics will face this economic reality on a much larger scale in the years ahead.

Billings added that some states have added to the subset of detected conditions to develop the information base for possible future testing. And Ramsey pointed out that the value of this information could turn out to be "extraordinarily high" and could be provided without incurring any additional immediate costs. "We could imagine a scenario that before [a test is added], we look at the value that that information would provide, the benefits and costs, before we allow [a test to be performed and results] provided as information to the patients," said Ramsey.

In relation to the limitations of self-assessment of ethnicity, one participant suggested that whole genome sequencing could be used to make estimations of biogeographical ancestry so that clinicians would have supplemental information. For example, they could recommend enzyme testing.

In response to a question about malpractice litigation arising out of genomic testing, Beaudet said that the possibility exists. For example, if a patient were offered a limited genetic test when a more comprehensive test would have uncovered a serious mutation, could that be considered malpractice?

5

Unprovoked Deep Vein Thrombosis

Important Points Made by the Individual Speakers

- Incidental findings in whole genome sequences that are actionable can be a source of value rather than a liability.
- A range of options for providing different levels of sequence information exist, and all can provide benefits to patients.
- A lack of epidemiological information is often more of a factor than economic uncertainties in cost-effectiveness analyses of genomic screening.
- In receiving the results of a genetic or genomic test, patients tend not to learn about the potentially harmful effects that a test result can have.
- The expertise of physicians who specialize in particular areas and are highly qualified will continue to be an essential part of genomic screening systems.
- Much more genomic data on different racial and ethnic groups are needed.
- The costs of interpretation and delivery of information to patients need to be decreased in order to ensure equitable access to genomic technologies.

In the second scenario discussed at the workshop, the woman who underwent preimplantation screening in the first scenario has developed a health problem 5 years later:

> The individual is seen at 40 years of age with progressive left lower extremity swelling and pain. Evaluation reveals an unprovoked deep vein thrombosis in her left lower extremity. She will be treated as an outpatient with low-molecular-weight heparin and warfarin. Targeted testing includes CYP2C9 and VKORC gene analysis.

Deep vein thrombosis is a common condition, said Frederick Chen of the University of Washington, when he introduced the case. About a half-million patients per year in the United States get a blood clot in the leg. About one-quarter of them end up with a clot that travels to the lung, causing a pulmonary embolism (Beckman et al., 2010). Clinicians are well trained and accustomed to dealing with this condition, which he suggested may change the bar for a new technology or medication to be used in treating a thromboembolism.

A CLINICIAN'S PERSPECTIVE

Various risk factors are associated with deep vein thrombosis in 40-year-old women, said Michael Murray, clinical chief in the Genetics Division of the Department of Medicine at Brigham and Women's Hospital. These factors include smoking, pregnancy, immobility, extended travel, surgery, hypertension, obesity, and cancer. Genetic factors may also be involved, including the prothrombin mutation and Factor V Leiden.

Warfarin is thought to impede the synthesis of clotting factors, specifically through inhibition of the vitamin K epoxide reductase complex C1 subunit (VKORC1). Therapeutic use effectively lowers the amount of active vitamin K–dependent clotting factor by approximately 30 to 50 percent. The effects of anticoagulation therapy generally occur within 24 hours, with peak effect taking up to 96 hours.

As a result, clinicians give doses of warfarin and then have to wait 2 to 3 days to determine the effect. The goal is to achieve an international normalized ratio (INR) between 2 and 3 to minimize the risk of either clot complications or bleed complications. "It is a very unwieldy tool," said Murray.

The U.S. Food and Drug Administration (FDA) label for warfarin explains its pharmacogenomics (Bristol-Myers Squibb, 2010):

> A meta-analysis of 9 qualified studies including 2775 patients (99% Caucasian) was performed to examine the clinical outcomes associated with CYP2C9 gene variants in warfarin-treated patients [Sanderson et al.,

2005]. In this meta-analysis, 3 studies assessed bleeding risks and 8 studies assessed daily dose requirements. The analysis suggested an increased bleeding risk for patients carrying either the CYP2C9*2 or CYP2C9*3 alleles. Patients carrying at least one copy of the CYP2C9*2 allele required a mean daily warfarin dose that was 17% less than the mean daily dose for patients homozygous for the CYP2C9*1 allele. For patients carrying at least one copy of the CYP2C9*3 allele, the mean daily warfarin dose was 37% less than the mean daily dose for patients homozygous for the CYP2C9*1 allele.

In an observational study, the risk of achieving INR >3 during the first 3 weeks of warfarin therapy was determined in 219 Swedish patients retrospectively grouped by CYP2C9 genotype. The relative risk of overanticoagulation as measured by INR >3 during the first 2 weeks of therapy was approximately doubled for those patients classified as *2 or *3 compared to patients who were homozygous for the *1 allele [Lindh et al., 2005].

Certain single nucleotide polymorphisms in the VKORC1 gene (especially the 1639G>A allele) have been associated with lower dose requirements for warfarin. In 201 Caucasian patients treated with stable warfarin doses, genetic variations in the VKORC1 gene were associated with lower warfarin doses. In this study, about 30% of the variance in warfarin dose could be attributed to variations in the VKORC1 gene alone; about 40% of the variance in warfarin dose could be attributed to variations in VKORC1 and CYP2C9 genes combined [Wadelius et al., 2005]. About 55% of the variability in warfarin dose could be explained by the combination of VKORC1 and CYP2C9 genotypes, age, height, body weight, interacting drugs, and indication for warfarin therapy in Caucasian patients [Wadelius et al., 2005]. Similar observations have been reported in Asian patients [Takahashi et al., 2006; Veenstra et al., 2005].

Murray estimated that 1 percent of the U.S. population takes warfarin, with a dose range between 1 milligram and 20 milligrams per day. Providers and patients try to stay between that INR value of 2 and 3 for 3 to 6 months and often longer. "You can imagine what a struggle that is," Murray said.

Though the FDA label provides guidance for doses depending on the CYP2C9 and VKORC1 genotypes, the decision memo from the Centers for Medicare & Medicaid Services (CMS) for pharmacogenomic testing for warfarin response states that testing of the variants will not be covered unless the patient is in a trial developing the evidence base for the use of the test. This is an example where medical science and medical practice do not correspond, said Murray. This is one of the drivers for why genotyping to predict warfarin dosing is not being done.

Murray briefly described the Medco-Mayo Warfarin Effectiveness Study (Epstein et al., 2010), which found that warfarin genotyping reduces hospitalization rates. An unusual feature of this study was that it did not provide genotype information to physicians before making the first dose decision. Rather, it provided that information about a month later, with 70 percent of physicians being told that a dose might be too high or low and that increased monitoring was warranted. Murray warned that the reduced hospitalization might be an instance of the "Hawthorne effect"—in which changes in patients' behaviors and health outcomes are related to the special treatment they received—and the results have not yet been replicated. The National Institutes of Health (NIH) was conducting a major study called the Clarification of Optimal Anticoagulation through Genetics trial at the time of the workshop to examine genotype-guided dosing.

Warfarin is a cheap drug, but a large infrastructure has been built up around warfarin care involving both clinics and home-monitoring structures. "Patients stay on target with their therapy better when they are monitored by experts," said Murray, which is "part of the cost of warfarin care."

Regarding the patient in the scenario, Murray said that he, too, would encourage her to stop smoking. The prothrombin gene mutation would give insight into predisposition but not have any specific management implications at the time of care. The variations that Murray assumed the patient had in CYP2C9 and VKORC1 would lead to lower initial dosing of warfarin, with the expectation of a lower daily dose over time. Nevertheless, she would still need to be monitored, and her therapy would still need to be adjusted. Because the deep vein thromboembolism was unprovoked, she would receive at least 6 months of warfarin and probably more. Murray also noted that at the time of the workshop, genotypic information would not be routinely available to help providers in making their decisions.

In terms of receiving additional information from whole genome sequencing, Murray observed that a fair number of people with unprovoked deep vein thrombosis have an underlying cancer. If whole genome sequencing were to reveal that she has a syndromic cancer risk, that information might be valuable in this scenario. Also, some of the other cytochrome genes are relevant in the metabolism of warfarin, and variants in those genes might play a role in deciding on a dose. Murray noted, however, that he would not have use for further incidental findings in managing this patient's acute case and suggested that conversations about ancillary information would be deferred to a later point in time.

Wylie Burke, professor and chair of the Department of Bioethics and Humanities at the University of Washington in Seattle, pointed out that costs will be associated with incidental findings from genomic screening. That information needs to be retained in a person's medical record for future reference and action, even if no action is taken at the time the finding is made.

A FUTURIST'S PERSPECTIVE

Euan Ashley, director of the Center for Inherited Cardiovascular Disease at the Stanford University School of Medicine, described his evaluation of a colleague's genome that had been sequenced as part of the Personal Genome Project (Ball et al., 2012). The colleague, Steven Quake, had asked Ashley about a variant in the myosin binding protein C gene, a gene that can be involved in sudden death in young people from hypertrophic cardiomyopathy (Maron et al., 2012). Ashley asked Quake whether anyone in his family had ever died suddenly at a young age, and Quake mentioned a cousin's son who had. "With that, he became my patient, and somewhat inadvertently I became one of the first physicians to have [access to his patient's] whole genome sequence."

Ashley gathered several other colleagues from Stanford, and together they performed a thorough interpretation of Quake's genome (Ashley et al., 2010). Shortly thereafter, Ashley was contacted by John West, chief executive officer of Personalis, Inc., who had recently had his own genome sequenced along with the genomes of several family members. (West describes his experiences in the next section of this chapter.) West had a family history of venous thromboembolism and had experienced a pulmonary embolism himself. Even after being put on warfarin, he had an unprovoked second pulmonary embolism. West was particularly interested in seeing whether he had a genetic condition that had been passed on to his children.

Ashley and his colleagues first determined whether West was a heterozygote for the Factor V Leiden mutation. This mutation is actually contained within the haploid human genome reference sequence because one of the anonymous contributors to the original sequencing project had the mutation. To cope with this and other deleterious variants that are contained within the reference genome, Ashley and his colleagues developed a synthetic reference sequence that contains the major population-specific allele at every position. They also used a newly developed technique to reduce sequencing error rates by up to 90 percent (Roach et al., 2010). These advances allowed them to build a robust platform from which to engage in genome interpretation.

Using these techniques, the Stanford group was able to deliver to West a list of potential risk alleles for a number of conditions that were shared among West, his wife, and their son and daughter (Dewey et al., 2011). The Factor V Leiden mutation had been passed from father to daughter, and other risk alleles had passed variously from the father and mother to the son and daughter. The group also put together assessments of onset, severity, actionability, lifetime risk, and variant pathogenicity for each risk allele, rating each factor between 1 and 7. "These are very arbitrary numbers, but we had to start somewhere," said Ashley.

"The genome has arrived," Ashley concluded. The task ahead is to learn what to do with it.

A PATIENT'S PERSPECTIVE

Finally, John West provided his perspective on the genetic odyssey he and his family had taken. He had an unprovoked pulmonary embolism at age 43; the median age for such an event is 60 (Silverstein et al., 1998). After checking into the emergency room, he was told that his condition was life-threatening. "People were concerned about me even sitting up in bed, that this would dislodge the clot and could cause more serious complications," he recalled. In the next 4 days, the hospital spent $22,000 on his tests and care. He was started on standard doses of heparin and warfarin without prior genetic testing. It was only after West was released from the hospital that he received results confirming he had a heterozygous mutation for Factor V Leiden.

His warfarin dosing turned out to be unstable even with very careful dietary restrictions and monitoring of INR. Six months later, West had a second pulmonary embolism. This incident led to suspicion that an occult cancer was causing the thrombophilia, though even after many additional tests, none was ever found. The warfarin dose gradually stabilized, however, and has been constant for more than 8 years.

On discussing his condition with his family, he found that his mother had been hospitalized for clotting in her legs when she was 40. The cause was unknown at the time, and the episode did not lead to any later screening in her children.

West had been involved with automated DNA sequencing since 1982, so he was an enthusiast for its use. When the company 23andMe started offering genotyping in 2007, he and his family had their genotyping done. The Factor V mutation was found in his mother and his daughter but not in his wife. But that test did not look at any other loci on the Factor V gene and did not assess structural variation.

In 2009 he, his wife, and their children had their whole genomes sequenced. The family then analyzed the sequences themselves, working with university groups. They found 13 other mutations in the Factor V gene. West's daughter was found to have inherited four nonsynonymous variants from his wife along with his Factor V Leiden variant. Further analysis of the four variants in his wife revealed that they were probably benign. No structural variants were found in the gene.

All this work was done on a direct-to-consumer basis, said West, "not because we wanted to work outside the medical system but because you couldn't do it inside the medical system."

When West's daughter turned 18, she was prescribed estrogen-based contraceptives to treat her acne. His daughter refused her dermatologist's prescription because she knew her genotypic information and the significantly increased risk of developing deep vein thrombosis and pulmonary emboli (Vandenbroucke et al., 2001). "Had she not had that information—and we got a lot of grief about having children sequenced—she would have been prescribed something that would probably have increased her chances of blood clots by 10 to 20 times," West said. Other precautions that West and his daughter take include using Tylenol rather than aspirin; avoiding foods high in vitamin K, such as spinach and miso; and avoiding injuries that could provoke internal bleeding. "This is not a big disaster in my life," he said. West maintains careful compliance with his warfarin dosing and has monthly INR tests to confirm the coagulation results. He noted, "This is an example where there is a genetic test where there is, in fact, a great deal that is actionable."

West addressed the question of the "incidentalome," or information from the genome that was not being sought and is not necessarily wanted. He described this issue as misplaced. The biggest medical issue he has had in his life is that the genetic testing was done after the deep vein thrombosis occurred. "You need to do the testing ahead of time. I do not want to be in that hospital, being told that I shouldn't move because the clot might get dislodged. I do not want to go through 8 months of testing to try to find a cancer that apparently wasn't there because people hadn't really looked at my genome." In his case, the incidentalome would have included the Factor V Leiden variation and other variants that could have led to practical and inexpensive lifestyle changes. "If I can make these kinds of changes and avoid the deep vein thrombosis and pulmonary embolism, that is where I want to be. And I think that is probably where we should explore what are the economic balances of heading in that direction," he added.

Information on the genome can be delivered at a wide range of price points, he observed. Whole genome sequencing is still expensive, but panels that cover many variants are much less expensive. Even though people have different financial situations and insurance coverage, a range of options could deliver substantial benefits. West said that he supports consumer choice and thinks that genome sequencing should occur within the context of medical practice, yet consumers who are interested in these options often have to go around the medical system "because the medical system has not dealt with this."

West also issued several cautionary notes. One is that genome sequences are not as accurate as people might think. False positive and false negative rates on variant detection are high and do not appear to be declining.

The technology has been improving, but the remaining errors are largely systematic, so accuracy has not dramatically improved.

Second, as with the Factor V variant, the human reference sequence contains many disease-related alleles. Also, many public databases have no mechanism to remove old data, and the errors in those databases have not been corrected, said West. Finally, no system exists to combine risks when multiple variants are known to predispose to a disease.

All these problems are solvable, West said. But he reiterated that many challenges remain in the technology arena as well as in the policy arena.

Finally, West made the point that the era of genomic testing need not cost the health care system a lot more money. If genomic screening were an add-on insurance option, like dental insurance, he would opt for it, have his family tested, and respond appropriately to any findings in which there was reasonable confidence and something actionable that could be done. "We haven't found that there is such a lack of actionable things that we have to delve into all the uncertain and vague results."

As more people learn about what genomic testing can offer, they will demand that the nation take advantage of the benefits to be gained, West said. "If we can solve some of these diseases and get rid of the actual burden of disease, we will have a much bigger impact on the economics of health than we will by rearranging different parts of the insurance system."

ECONOMIC PERSPECTIVES

Veenstra began his presentation by noting that the development of genomics will have an effect on the U.S. economy as a whole. It will influence job growth, the formation of companies, reimbursement policies, and health care policies in general. But the focus of the workshop, he noted, was on the value that genomics could bring to the health care system, patients, and society. Only after this value is understood can other policies be generated.

To determine the incremental value of genomic testing, researchers need to compare the use of the test with a different course of action, said Veenstra. For example, in the case of warfarin treatment, is testing being compared with treatment without testing or with no treatment at all? If testing is performed, when do the provider and patient receive the test results? What are the consequences of those results, both for future treatments and for other decisions that a provider or patient might make?

Furthermore, economic analyses need to take all of the possible consequences into account. For example, warfarin treatment is designed to mediate between the risk of bleeding and the risk of clotting in a patient. Either of those outcomes could have severe consequences for a patient, including death and long-term disability. "When you are working through an economic evaluation, you look at that decision, and you think about every single thing that could happen to that patient. And we want to try to include all of those aspects as best we can," Veenstra said.

In the current scenario, a CEA would take into account the costs of the tests, continued monitoring, the drug, any adverse clinical outcomes, and other factors. An outcome of this analysis might then be the cost per clot avoided or cost per life year saved. A CUA then would add an assessment of the patient's quality of life, such as whether a patient could have a long-term debilitating outcome. This latter technique, which often measures patient outcomes in terms of QALYs, needs to account for patient preferences—for example, by conducting surveys that ask people to rate their health state if they were debilitated by a stroke—though not necessarily for whether they get genetic testing. CUAs are preferred from a theoretical perspective, observed Veenstra. Still, real-world decisions are often more influenced by CEA studies.

Grosse pointed out that the weakest links in CEAs usually involve epidemiology and clinical effectiveness rather than economics. Does an intervention actually make a difference in terms of the health outcomes? "If we don't have evidence of effectiveness, we do not have cost-effectiveness," he noted.

Ashley made the interesting point that relatively little evidence exists to guide treatment decisions for most of the patients he sees, even though his specialty, cardiology, has better clinical trial evidence than most fields of medicine. But these clinical trials have very tight inclusion criteria for patients so that statistically significant results can be obtained when examining small effects, and most of Ashley's patients would not qualify for the trials and are therefore not necessarily described by the trials' results. "The reality of medicine as we practice it is that we don't have evidence for most of the things that we do."

Grosse acknowledged that diverse sources of evidence, not just the results of clinical trials, are necessary to make conclusions about cost-effectiveness. But these varied sources of evidence might still not be sufficient to translate into a recommendation for coverage or clinical practice.

Veenstra agreed that evidence is a key component of being able to make a health economic assessment and that for genomics a real-world standard is being applied for how much evidence is needed. He noted, however, that stakeholders are not aligned on that standard and that the varying thresholds impact real-world policy and decision making. The question becomes, said Ramsey, "how comfortable are you with the chance that you might be making a mistake," whether it be in providing a test that turns out to have no benefit or withholding a test that turns out to have great benefit.

One participant pointed out that medicine is much more sensitive to errors of commission than errors of omission. In other words, medicine is much more willing to risk harm by not providing something than by doing so. Patients are not happier to die of a clot than bleeding, he said. Some of this bias results from concerns about liability, but the harms can be equivalent. "We need to be honest about assessing the whole landscape," he said.

Also, the chain of evidence between a decision and an outcome can be long and complex, the participant continued. Most anticoagulation outcomes are secondary outcomes, such as time in range or time to stable dosing as opposed to direct outcomes of bleeding and clotting. The question is how much confidence there is that those secondary outcomes are predictive of primary outcomes. Intermediate end points such as genetic testing may be able to help bolster that confidence by assessing risk.

The participant also noted that West's family history already captured the risk of thrombophilia regardless of the status of his Factor V Leiden gene. Not using this family history represents an opportunity cost that needs to be recognized. "If we are spending our money on genomics, then that means we are ignoring other things. And in the realm of anticoagulation, if we focus on genomics, are we ignoring the opportunities to do other things like clinical decision support with guided dosing," he asked. Similarly, is genotyping misplaced in urban settings where there are anticoagulation clinics?

Decision points can differ on the basis of the available evidence, Grosse said. For example, considering an option has a lower threshold of evidence than deciding to choose an option. Grosse also noted that clinicians will continue to make judgments about treatments for their individual patients regardless of standards that are set across a population, given the heterogeneity of preferences and treatment effects. "It is not one-size-fits-all."

Ramsey said that the best approach would be to have a cost-effectiveness study for the average population and then to modify those results on the basis of the characteristics of each patient. A system can also be structured to encourage the more cost-effective approaches and discourage the cost-ineffective practices, with flexibility for individual decisions on the basis of patient characteristics.

Veenstra said that changing clinical management poses its own risks. For example, monitoring of phenotypes, as with INR measurements, works fairly well, and genomic-based treatment decisions need to be compared with that standard.

Finally, Ramsey mentioned that patients may not fully comprehend the impact of receiving a test result. They may test positive and be relieved that the condition was caught in time and they can be treated or they may test negative and feel they do not have to worry about developing a genetic condition. But patients tend not to learn about the untoward effects that a test can have, such as distress or anxiety to patients and their families, the development of a false sense of security regarding risk of disease, results being uninformative for decision making, the potential need for additional confirmatory testing for positive tests, or actual harm from unnecessary procedures based on a false positive result. "It leaves those of us on the front lines in a real dilemma, particularly when there are advocates who want that test and there is evidence that does not support that advocacy," he said.

DISCUSSION

In response to a question, several panelists discussed the related topics of access and equity with regard to genomic technologies. In the past, investigators have often experimented on themselves or their colleagues first, because those individuals can give well-informed consent. But that approach focuses on justice as protection from research harms, not on justice as access.

Ashley observed that the analyses of both Quake's and West's genomes fell into the category of research as opposed to health care. These individuals were chosen first because both understood the potential benefits and harms of whole genome sequencing. Equity will be important moving forward from this point, he said. Ashley also noted that he participates in a free clinic once a month so that cardiology patients who cannot come to his clinic at Stanford can see him in an alternate setting, and "I would be as willing to entertain the idea of whole genome sequencing in that setting as I would in the clinic at Stanford."

West pointed out that when he approached Ashley's group, he and his family had already had their genomes sequenced. He also stated that the largest costs in the future will not be the generation of a sequence but the interpretation of that sequence and communicating that information to patients. If those costs stay high, the health care system will ration the use of the technology in some way. To provide more equitable access, researchers will need to reduce these costs, whether through Web-based tools or other means.

West was asked whether he was concerned that some of the actions he took in response to his genetic results were not based on solid evidence. West responded that he and his family sought to find the best evidence they could. They relied on advice from physicians who specialize in these areas and are highly qualified. "I expect the physician to have the judgment to know what the recommendations are. And those may be based on their personal experience in some cases," he said. Billings agreed, pointing out that this professional expertise is always going to be a necessary part of the system and will need to be taken into account in projecting costs.

Evans observed that West's Factor V genetic variant is a risk factor for deep vein thrombosis, but it is a modest risk factor. For example, discouraging women from taking birth control pills who are heterozygous for Factor V Leiden would create many more medical complications from unwanted pregnancies than it would prevent clotting problems. "We have to think very hard before we say that we are going to tell millions of people that they are at a high risk for clotting, when the reality is that it is a very modest risk factor," he said. In addition, as Calonge pointed out, false positives will be generated by these genetic tests in a pre-event prevention setting, and harms could be associated with those mistaken test results. He also noted that, for most individuals, just knowing you are at increased risk for a disease is insufficient to change behavior and suggested that West's family may not be representative of the average population in adopting changes.

Another topic of discussion was the focus to date of genomic data on some populations and not others. A number of panelists agreed that much more genomic data on different racial and ethnic groups are needed and that caution should be used when interpreting information for different groups. Ashley pointed out that databases of single nucleotide polymorphisms are being developed that are much more heterogeneous in terms of populations studied.

6

Cancer Care

> **Important Points Made by the Individual Speakers**
> - Molecular characterization of drugs has far preceded the molecular characterization of cancer.
> - As cancer drugs target smaller subsets of people, new drug development paradigms are needed that require smaller trials and demonstrate substantial effects.
> - Regulatory science needs to advance to be able to approve effective drugs with limited amounts of information.
> - Despite the value that genetic testing can produce, a relatively small number of patients with cancer receive such tests.

In the third scenario discussed at the workshop, the patient has developed cancer 5 years following her deep vein thrombosis:

> The individual is seen at age 50 with cough, dyspnea, and chest discomfort. Evaluation reveals a lung mass; bronchoscopy and biopsy reveal a non-small-cell lung cancer. Her tumor is found to have variations that allow the use of targeted therapy, and with treatment the patient goes into remission.

A CLINICIAN'S PERSPECTIVE

The clinician who discussed the case, Kenneth Offit, chief of clinical genetics service in the Department of Human Genetics, Memorial Sloan-Kettering Cancer Center, changed the case slightly to make it more realistic. He posited that the woman had advanced non-small-cell lung cancer, because if it were a local lung cancer she would simply be referred for surgery. He added that the patient experienced a relapse after going into remission and that the case was followed by further testing and treatment.

Standard chemotherapy for advanced non-small-cell lung cancer, as that treatment existed about a decade ago, produced only a median survival increase of between 7 and 8 months, Offit observed (Kelly et al., 2001; Schiller et al., 2002). Even when tyrosine kinase inhibitors began to be used in 2003, prior to FDA approval of epidermal growth factor receptor (EGFR) mutation testing, they produced only modest survival advantages (Kris et al., 2003). Once testing for EGFR gene mutations began in 2004, response rates reached the remarkable rate of 75 to 82 percent, said Offit, with progression-free survivals of 8 months to a year or more (Pao et al., 2004).

Nevertheless, most cases are not success stories. And resistance to tyrosine kinase inhibitors commonly develops (Balak et al., 2006; Bean et al., 2008), which requires treatment with another targeted drug. Randomized controlled trials for EGFR testing have been carried out and point to a progression-free survival on the order of 3 to 4 months (Mok et al., 2009). "It is a striking advance, but we want to keep all of this in perspective," Offit said.

Current practice at Sloan-Kettering is to test lung cancers with targeted testing. Driver mutations for between half and two-thirds of lung cancers have been identified and can be tested, and these mutations point to targeted therapies that have already been approved by the FDA or are in the pipeline for approval. Still, Offit noted, access to testing can be a problem, depending on geographical residence.

Offit briefly described several studies that have found costs per QALY of around $100,000 for targeted cancer interventions (Atherly and Camidge, 2012; Bradbury et al., 2010). These costs per QALY are higher than for BRCA1 testing (Plevritis et al., 2006) and testing for Lynch syndrome (Ladabaum et al., 2011).

In discussing the second model of genomic information delivery in which targeted data plus other actionable information is provided, Offit described the Integrated Mutation Profiling of Actionable Cancer Targets screening model, which captures information on 230 cancer genes, though Sloan-Kettering is currently running this as a research arm only. Two major issues with this type of screening are variants of unknown significance and the required consents. These variants can impact pharmacogenomics as well as noncancer disease risk in future generations.

Offit noted that they spend 2.5 hours to do the counseling for next generation sequencing panels and wondered what could realistically be done when potentially receiving information from 20,000 genes. And he noted that companies who provide these services use consent forms that are impractical for patients. "Your patient is supposed to say, 'I am in the mood today to look at my mitochondrial genes, but let's leave out the PI3 kinase pathway, okay?' And they will write that down on the commercial consent form. Rather unrealistic."

As an alternative, he pointed to a consent form developed at the University of Michigan that asks whether patients want to receive results "that do not have a direct impact on [the] care of my current cancer." Patients can choose to receive results "that may have significance for biological family members" or results "that are not related to your cancer but may have potential medical impact for you."

Finally, Offit briefly mentioned the delivery of complete information from whole genome sequencing, including the incidentalome. In cancer therapy, whole genome analysis may reveal germline changes that impact therapeutic activity and that have implications for familial risk as well. Full disclosure of this information without medical practice standards, he said, will likely lead to clinical and economic inefficiencies—if not to chaos. "We have a lot of cheap tests in medicine. But just because it is cheap doesn't mean that you take it outside of the medical model," he concluded.

A FUTURIST'S PERSPECTIVE

The best way to predict the future is to understand the past and where we are now, said Stephen Eck, vice president and global head of medical oncology for Astellas Pharma Global Development. In the past, the pharmaceutical industry was not very interested in oncology. Drug discovery was highly empiric, drawing extensively from natural products and alkylating agents. Companies could sell only a few doses to any one person because of cumulative toxicity and low incidence. "There were better things to do with your capital if you were a drug developer," Eck said.

This situation began to change as companies recognized that a drug used for a few adults in one cancer market could also be used in other cancer markets, said Eck. Drugs became more tolerable and could be used longer, which increased sales. Premium pricing made small single indications attractive. Scientific advances made rational drug discovery faster and cheaper. For all these reasons, treatments for small subsets of cancer became more economically attractive.

Today, a variety of drugs are used to treat non-small-cell lung cancer.

Most of these drugs are used without any genetic testing, and the market is highly segmented, with people making decisions using data of varying quality. The best situation would be for the market to segment along scientific lines so that patients received what they needed, not along the lines of what people thought on average would be a good idea, Eck said.

Molecular characterization of drugs has far preceded the molecular characterization of cancer, Eck observed. Today, drugs are highly purified and have well-known structures tied to their pharmacology. "We can understand the drug itself at an exquisitely detailed level," said Eck. Diseases such as lung cancer are much less well understood, but molecular diagnostics are starting to reveal their secrets. In the future, said Eck, "somebody will be diagnosed with lung cancer. They will get a biopsy. It will be molecularly characterized. And the therapy will be chosen based on the unique attributes of that patient's tumor." These therapies are not curative, and as tumors mutate and metastasize, new therapies will be needed. But cancers can be periodically reassessed, leading to selection of the next targeted therapy.

For this vision to become economically feasible, several things are needed, said Eck. Faster and cheaper technologies for molecular characterization need to be developed. New drug development paradigms are needed that require smaller trials and demonstrate substantial effects. Smaller trials will require improvements in pharmacovigilance, because clinical and biological information will need to be collected once a drug is put out into the marketplace to substantiate trial results. Regulatory science needs advances to be able to approve effective drugs with limited amounts of information. In return, Eck said, drugs need to be used effectively in oncology, especially with so many being used off label for nonapproved indications with little to no evidence to support that use. To pay for schools, food, and other basic services, we must end the steady increases in the cost of health care, said Eck, and drug development could help control costs if it could bring drugs to market for less money.

A PATIENT'S PERSPECTIVE

In 2004, at the age of 44, Richard Heimler was returning home from a business trip when he began experiencing chest discomfort and shortness of breath. A chest X-ray taken in the emergency room revealed a 3-millimeter spot on his right lung.

A positron emission tomography scan and biopsy confirmed that he had a malignant non-small-cell lung cancer, even though he was not a smoker and had no family history of lung cancer. "At that point, after getting that diagnosis, I was numb and naïve," Heimler said. "I was glad at the time I didn't know that 60 percent of lung cancer patients die in the first year and that 90 percent die in the first 5 years. I also did not know that 200,000 people are diagnosed with lung cancer every year, and 180,000 people die from lung cancer every year" (ACS, 2012).

Heimler was fortunate that his mother was a genetic counselor. Serving as his advocate, she found him the best doctors, hospitals, and treatments in New York. "I recommend that everyone has an advocate, or someone to do the research, absorb the information, or at times fight for you. Newly diagnosed patients and their caregivers should know their patient advocacy organizations. They are a great resource and support," Heimler said.

He initially had a pneumonectomy—the removal of his right lung—followed by chemotherapy. He then had a recurrence of the cancer, this time a brain tumor, which was removed surgically. A later brain tumor was killed by gamma knife radiation. He then had a tumor below his rib cage, which was removed surgically, followed by chemotherapy. "I know this sounds like a horror story, but I am still standing," he said. Then, 3 years ago, spots appeared on his left lung. "That was our worst fear, because that's the only lung I have. So at that point, if it's a baseball game, I was in the bottom of the seventh inning, top of the eighth."

At that point, Heimler's oncologist tested his tissue to identify any tumor variations associated with his cancer. He tested positive for the ALK gene, which is only present in 3 to 5 percent of patients with lung cancer (Kris et al., 2003; Riess and Wakelee, 2012), and his oncologist told him about a new clinical trial with Pfizer for patients with this tumor variation. Drug companies and diagnostic testing companies need "to educate our doctors about all the new targeted treatments," Heimler said. "Doctors must recommend to all their cancer patients to have diagnostic testing for all available targeted treatments. It is also important to implore our doctors to archive enough of our tissue to get an accurate sample for diagnosis."

Because the tumors on his lung were small, the risks were minimal for Heimler to enter the stage 2 clinical trial for 6 months. He began taking the drug crizotinib (Xalkori)—three 250-milligram pills in the morning and another three pills in the evening—and has experienced minimal side effects.

A few months after beginning the trial, his tumors began to shrink and then cavitate, and no new activity occurred. The week of the workshop was his 2-year anniversary on crizotinib. He reported, "My doctors cannot see any tumors, and I feel great and have a good quality of life. So I am very, very thankful."

Heimler has found the clinical trial to be economically advantageous. His treatments are free for the rest of his life as long as he stays in the trial, as are his computed tomography scans, the magnetic resonance imaging scans of his brain, and other tests. If he leaves the trial, however, his drug costs $6,000 every 3 weeks, of which Medicare covers only about $4,000.

He does not qualify for any financial assistance for his medication, but his parents are helping to ensure that economics is not a factor in his care. He has two private disability plans, plus Social Security disability. "I will never be rich, and I will never be able to save any money, but I have sufficient income to live comfortably and raise my two children."

Lung cancer may not be curable, but it is treatable and livable, Heimler said. "Lung cancer is a relentless, unforgiving, and nondiscriminating disease. But thanks to advances in personalized medicine and diagnostics, it does not have to be a death sentence." On the basis of his experiences, he has become an advocate devoted to raising awareness for lung cancer. Lung cancer is the number one cause of cancer deaths in the world. It kills more people than any disease but heart disease, and it kills more people than breast, colon, kidney, skin, and prostate cancer combined (ACS, 2012). People with lung cancer are often subject to a "blame the victim" mentality, because other people think they brought the disease on themselves by smoking. This perception has an effect on research dollars, Heimler observed. According to the Lung Cancer Alliance, federal research funding per cancer death is $26,398 for breast cancers, $13,419 for prostate cancer, $6,849 for colon cancer, and just $1,442 for lung cancer.[1]

"I hope that in the years ahead we can change the perception of lung cancer in the general public, the medical profession, the government, and the media in order to generate a fair and proportionate share of research funding for this deadly disease. We need to raise lung cancer to a national public health priority," said Heimler.

"I am one face of lung cancer," he continued, "who is benefiting from diagnostic companies and drug companies working together to identify and produce targeted therapies to make a difference for lung cancer. And even though my targeted therapy might only be effective for 3 to 5 percent of lung cancer patients, there are still 11,000 people each year who can benefit from this drug alone. It is time to take advantage of advances in personalized medicine by getting the right medicines to the right people. This new world of science is giving me hope that maybe my children and I will have more time to create new memories in the years ahead."

ECONOMIC PERSPECTIVES

The panel of economists began by discussing an observation made by Heimler and several other workshop participants—that genetic testing still occurs for a relatively small number of cancer patients. As Ramsey observed,

---

[1] According to estimates by the NIH, fiscal year 2011 funding for lung cancer was $221 million, breast cancer was $715 million, colorectal cancer was $313 million, and prostate cancer was $284 million (http://report.nih.gov/categorical_spending.aspx [accessed March 19, 2013]).

genomic testing can provide enhanced value if patients are directed to drugs where they have a better response rate and fewer side effects. But the delivery and reimbursement systems are structured in a way that acts to restrict rather than enhance testing, which may be a hindrance to rapidly evolving personalized medicine.

Billings pointed out that costs are involved with moving samples among sites and conducting different tests even as a disease continues to progress. One clear benefit of a rapid and comprehensive mutational test is that it would produce comprehensive knowledge quickly.

Veenstra commented that somatic genomic testing is quite different from the other scenarios examined at the workshop. In the case of cancer, the testing clearly has the potential to provide benefits to patients by avoiding the costs and adverse effects of ineffective treatments. Also, at least for some forms of testing, the cost of the test is much less than the cost of the therapy. "From a value perspective, why wouldn't you do [genetic testing]? What is hard to get my head around is that only a minority of patients are being offered this testing."

Heimler added that the first thing he asks people who have been diagnosed with lung cancer is whether they have been tested for the ALK gene. Mostly they say no because they are smokers and their doctors have told them that they probably do not have the gene. "But I think it is absurd for a doctor not to test every single patient, especially with a disease like lung cancer where this might be our only hope of extending our lives." Veenstra noted that clinicians are not used to treating patients in which just a small percentage of people respond to a specific medication. "It is a different paradigm," he said.

Eck responded that medicine is an inherently conservative enterprise that is slow to change. For example, the data in favor of lumpectomy versus radical mastectomy far outpaced its uptake in the community, but eventually it was adopted. One participant added that this type of slow adoption is observed with every type of new technology, not just genomics.

DISCUSSION

A participant observed that the increasingly complex landscape of cancer genomes poses an immense informatics challenge to be able to aggregate, analyze, interpret, store, and share these data. Many cancer centers have strong data systems, but they tend to be siloed, sometimes even within centers. Data from these systems need to be exchanged across institutions, across cities, and across the globe. In addition, the challenges in sharing data are cultural, financial, and technical, and all three of these areas must be addressed.

Timothy Ley, the Lewis T. and Rosalind B. Apple Chair in Oncology at Washington University, suggested that the research community is committed to sharing data on cancer genomes and somatic mutations. He mentioned that The Cancer Genome Atlas is helping to share data widely. Another participant pointed out that molecular diagnostics companies and the FDA both have significant roles to play as well. Evidence on the use of genomic tests is hard to generate without data being shared, and government agencies can help facilitate this sharing, the individual said.

Eck pointed to some of the difficulties in sharing data, including intellectual property provisions, restrictions in informed consent forms, and societal norms. Nonetheless, industry is becoming more transparent as it recognizes the advantages of increased access to information. And some companies have begun putting information immediately into the public domain because patenting can be costly and difficult, and releasing information prevents other companies from patenting it. "It is hard to protect data. It is expensive. It is sometimes just easier to give it away," Eck said.

One participant proposed an economic study that would compare targeted treatment based on EGFR testing with standard approaches to therapy as a way of demonstrating the value of testing. He also observed that cancer is undergoing a paradigm shift away from a histological definition of cancer toward a molecular characterization, which makes it difficult to compare new approaches with the current standard of care.

Offit also observed that society cannot afford to pay for very expensive cancer drugs that produce, on average, only a few extra months of life. Even for people whose expenses are covered because they are in clinical trials, those costs have to be covered somehow. Grosse agreed, noting that the common threshold of paying $100,000 per QALY is not affordable in the long run. "It works for rare diseases or for short-term treatment, but it can't go to scale."

Ramsey said that targeted therapy is probably not going to lower the cost of cancer care. Producing better results does not mean low drug costs. In fact, the drugs that elicit better responses tend to be more expensive. For drugs such as bevacizumab and crizotinib, with the current price structure and gains in survival, "it is not particularly good value."

In response to a question about why new drugs tend to be priced at about $100,000 per life year saved, Ramsey said that drug pricing is extraordinarily complex, "and I don't think there is always science involved." A drug may be developed for one cancer and produce a given survival gain, but when it is used for another cancer and produces a much smaller survival gain, the price is the same, even though the value of the drug differs from cancer to cancer.

Katrina Armstrong, professor of medicine at the University of Pennsylvania School of Medicine, noted that the way billing and reimbursement are done in hospitals can discourage testing. These financial systems should be set up so that tests are done at the time of maximum value. A participant noted that CMS may accept a proposal to move toward a billing system that relies on a test-specific code, which could ease this problem. The interpretation of a test result would be part of the physician's fee schedule, however, which may further influence incentives.

7

Panelists' and Stakeholders' Perspectives

> **Important Points Made by the Individual Speakers**
>
> - Genomics cannot afford a bubble of unwarranted enthusiasm.
> - Health economics research needs to consider both the decisions of individuals and the behaviors of populations.
> - A major hurdle to genomic-based medicine is the lack of a national, dynamically updated, interpretative database of evidence for the clinical utility of genetic variants.
> - Patients need time with their physicians to discuss options, but incentives are currently not aligned well to encourage this time.
> - If genomic technologies add to the cost of health care without commensurate benefits, they will not be widely adopted.
> - Data on the comparative effectiveness of genomic testing will take at least a decade to accumulate, which will delay the implementation of testing.

On the second day of the workshop, speakers reflected on the previous day's discussions. Although none of the presenters purported to speak for an entire sector of stakeholders, they made points that relate to many of the economic issues underlying the development and use of genomic tests.

A CLINICIAN'S PERSPECTIVE

James Evans, who provided the introductory presentation summarized in Chapter 2, emphasized that genomics cannot afford a bubble of unwarranted enthusiasm. When people engage in wishful thinking without seeking out evidence, they can overvalue any commodity, from real estate to tulips to genomic tests. Genomic-based medicine has great potential and is likely to produce great advances in reducing patient suffering and improving care, but evidence is needed to prove these points rather than assuming that these advances will occur.

He also emphasized the importance of focusing on big problems where the maximum returns in terms of outcomes and cost will occur. Accumulating evidence will point to where the greatest outcomes are likely to occur, and this evidence should be heeded.

Free medical tests do not exist, said Evans, and the inappropriate use of a test adds to the burden of health care costs. Everyone has an interest in seeing that tests are used wisely and appropriately, because everyone pays for those tests through insurance. Evans argued against the concept that personal utility should be a factor in deciding whether a genetic test should be used because "it can mean anything to anyone. Before I help pay for your personal utility quest, I want to see some evidence that it improves outcomes." Coverage decisions should be driven by commonly shared values, said Evans, rather than outliers. Determining and acting on these values is critical in many areas of genomic testing, whether informed consent, privacy, or cost. These decisions also need to maximize equity, though absolute equity is impossible. Existing medical tests are not used in what Evans called "shotgun or nontailored" ways, and genomic tests should not be treated differently.

Patients do not always choose to access all the information that is available to them. Many men and women who are informed about the benefits of particular screenings will still decide not to have the test. More information is not always better or desirable, said Evans, and extra information that is not of obvious value can be counterproductive.

All these points argue for the use of targeted genomic testing, Evans said, where particular information is gathered when it makes sense to do so. Targeted testing will also focus educational efforts so that people and providers can be educated in categorical ways rather than test by test.

Finally, Evans noted the tremendous obstacle of the cost of targeted drugs for small populations. Even if genomic testing can reveal the characteristics of a tumor, the cost of a drug may make treatment for that genetic lesion prohibitive.

A RESEARCHER'S PERSPECTIVE

Armstrong focused her comments on how resources are allocated within the health care system. For example, novel tests and treatments would have their biggest impact on common diseases, but they could also have a substantial effect on rare diseases if many different diseases could be treated. Decisions need to be made about how to allocate resources under such circumstances, and genomics itself can inform these decisions.

Much of the discussion at the workshop focused on decisions at the individual level, Armstrong observed. But that level represents only one end of a spectrum. Decisions range from the micro level of individual patients to the macro level of the health care system or payer. Health economics research should look across this spectrum and consider how a new technology can improve decision making and outcomes within large populations as well as in individual health decisions.

Armstrong also discussed several challenges at the intersection of genomic testing and economics. First, health economics research faces a workforce challenge in that many researchers have been working in the field for a long time. The next generation of researchers needs to be engaged to work on the major issues confronting the field.

Also, the field continues to struggle with data in that many data are lost or not available when needed. A system is needed to gather, store, and disseminate the data that will be needed to resolve major questions.

Finally, health economics research needs to be focused on helping to answer constructively whether a technology should be used, and resources need to be allocated for doing so, said Armstrong. For example, the recent focus on patient-centered outcomes provides an important forum for taking economic studies forward. The true impact of genomics may lie in preventing the use of unnecessary tests rather than in linking any particular single nucleotide polymorphism to a disease or in determining how a whole genome sequence is used, said Armstrong.

A CHIEF SCIENTIFIC OFFICER'S PERSPECTIVE

Whole genome or exome sequencing is initially entering clinical practice informally via academic medical centers and biotechnology laboratories operating under guidelines from the Clinical Laboratory Improvement Amendments (CLIA), observed Thomas White, retired chief scientific officer at Celera Corporation and regent's lecturer at the University of California, Berkeley. Most of these tests are not being reimbursed by insurance companies, but when they are, compensation is made because the tests are "under the radar" and not consuming enough resources to warrant a thorough review for reimbursement.

Some researchers have argued that all genomic tests should be approved by the FDA, but that scenario is extremely unlikely, White said. For example, the path to FDA approval for a whole genome sequencing instrument and reagent system is currently unclear. Such a system would have complex intended uses, accuracy problems, no gold standard for comparison, rapid technical obsolescence, and a potential requirement for lengthy and costly prospective treatment-by-genotype clinical outcome studies. In addition, it is difficult for manufacturers that have to produce a system with validated reagents and software to compete with laboratories that develop tests under CLIA that use research-use-only instruments and that can update software and reagents as needed. "There is an uneven playing field that will prevent any of these tests from reaching FDA-approved instrument systems for decades," said White.

A major hurdle to genomic-based medicine is the lack of a national, dynamically updated, interpretative database of evidence for the clinical utility of genetic variants, White said. Not every CLIA lab will be able to maintain and update a database of all its clinically useful variants and convey those updates to patients and physicians. Without evidence of clinical utility and cost-effectiveness, private and public payers may default to nonreimbursement. The government could have an extremely useful role in standardizing this information, said White, who suggested that the 100K Foodborne Pathogen Genome Project[1] may be a model to emulate.

The basis for reimbursement of whole genome sequencing remains uncertain, he observed. Except perhaps for cancer indications, single clinical indications may not be cost-effective, said White. Furthermore, because of issues with accuracy, targeted confirmatory testing of some actionable variants identified by whole genome sequencing may be necessary and will increase overall costs. Given these uncertainties, White asked, how can the cost-effectiveness of whole genome sequencing be assessed for application over a lifetime?

Professional organizations need to develop guidelines for reporting genetic variants to patients and providers, according to White. Also, providers and patients are going to want evidence that a test will make a difference in treatment.

Stakeholders may need to accept different evidentiary standards for the clinical utility of tests for different medical conditions, White said. Also, the government will need to continue to support prospective randomized clinical outcome studies, with studies prioritized by disease area. Studies supported by the Patient-Centered Outcomes Research Institute could contribute to this goal.

Cost-effectiveness studies are not terribly expensive, said White, and

---

[1] See http://100kgenome.vetmed.ucdavis.edu/index.cfm (accessed March 19, 2013).

information from previous randomized clinical outcome studies can be used to assess cost-effectiveness. For example, data from the Framingham heart study have been used to look at genetic markers that increase the risk of heart disease and the costs of preventing heart disease in various ways (Shiffman et al., 2012).

A PATIENT'S PERSPECTIVE

Patients tend to differ from patient advocates, said Mary Lou Smith, cofounder of the Research Advocacy Network. Patients are more focused on themselves, their treatment, their families, and their future, whereas advocates look at a broader picture with the needs of a composite patient in mind. Smith is a two-time breast cancer survivor and spoke as a patient, not as an advocate.

Patients want to know what the best treatments are for them. They want to know whether they will benefit from a treatment, suffer the effects of toxicity, or both. If a genomic test will provide that information, the question becomes whether a physician will order the test, whether the physician will understand the test results, and whether the physician will use the test correctly, said Smith.

Patients need time with their physicians to discuss options, but physicians do not have incentives to make this time. They are pressured by their administrators to be productive, and chemotherapy is profitable for hospitals, which can distort incentives throughout the system, said Smith.

Research is needed on how information from genomic tests is used by patients, said Smith. For example, studies conducted by the Research Advocacy Network indicate that benefit matters more than toxicity to patients when making treatment decisions. The greater the benefit, the more likely a patient is to choose treatment. The higher the toxicity, the less likely a patient will choose a treatment, although the nature, duration, and severity of the side effects are also factors. "It's not a simple process," she said.

One important application of genomic testing will be to identify patients who will not benefit from particular treatments. Such testing, Smith said, could produce both tremendous cost savings and great improvements in quality of life.

Smith also commented on informed consent. "I don't think anyone in this room actually believes those 27-page documents inform patients," she said. Even easy-to-read consent forms do not necessarily increase retention or understanding. "We need to start from square one and say, What does a patient really need to make an informed decision? And what is an informed decision?" Smith noted that she is a lawyer but still did not read the consent form given to her when she entered into a clinical trial. "I knew the people, I trusted the people, I knew the research they were doing, I wanted it to go forward."

She expressed concern about the delayed effects of treatment. If a woman develops a heart condition 10 years after receiving radiation treatments for cancer, there is no way to explore the connection between these two events because no system exists to make those linkages. And without knowing about such linkages, patients cannot be fully informed of the risks.

Smith emphasized the importance of engaging patients as partners if researchers want samples of tissue and data over time. This would provide a much fuller picture of the benefits and side effects of treatment.

Finally, evidence-based medicine requires new trial designs and end points for Phase I, Phase II, and Phase III trials, said Smith. "Right now, we don't have them."

A PUBLIC HEALTH OFFICER'S PERSPECTIVE

Public health thinks about the individual within the context of a community, a nation, and the global population, observed Calonge. Public health is more than publicly funded health, though that is part of it. Viewed broadly, public health provides the following essential services:

- Monitoring health status to identify and solve community health problems.
- Diagnosing and investigating health problems and health hazards in the community.
- Informing, educating, and empowering people about health issues.
- Mobilizing community partnerships and action to identify and solve health problems.
- Developing policies and plans that support individual and community health efforts.
- Enforcing laws and regulations that protect health and ensure safety.
- Linking people to needed personal health services and assuring the provision of health care when otherwise unavailable.
- Assuring a competent public and personal health care workforce.
- Evaluating effectiveness, accessibility, and quality of personal and population-based health services.
- Conducting research for new insights and innovative solutions to health problems.

Calonge focused on more of a macroeconomic evaluation of genomics and public health. He noted that in Colorado, premiums for state Medicaid medical services will exceed school finance's share of the general fund in about the year 2017. By 2019, Medicaid and K–12 education will consume the entire general fund, leaving no money for any other state services. "We have to get control of costs," he said.

If genomic medicine only adds cost, it will worsen rather than lessen this problem. Furthermore, genomic testing may not be covered by many of the insurance plans offered, said Calonge, even under health care reform.

Calonge also pointed out, as did James Evans (Chapter 2), that well more than half of all deaths result from causes that do not require genetic testing. Many are simply from old age. In Colorado, for example, of 30,000 deaths annually, 6,500 are from cardiovascular disease; 1,500 are from stroke; and 6,500 are from cancer, including 1,500 from lung cancer, 500 from colon cancer, 500 from breast cancer, and 40 from cervical cancer. Calonge estimated that more than half of these deaths could be prevented through efforts to control tobacco use, obesity, high cholesterol, and hypertension and by screening for colon, breast, and cervical cancer. Genomic testing could produce incremental improvements in health, but these improvements are much less beneficial than those available by applying what is already known.

Calonge concluded by pointing out that most genetics research is focused on discovery and the process of converting discovery to applications. Very little is devoted to the development of guidelines, the conversion of guidelines into practice, or converting practice into health impact in communities. Genomic medicine will fail if it rarely gets beyond discovery and applications to public health impact, Calonge said.

A HOSPITAL ADMINISTRATOR'S PERSPECTIVE

Hospital administrators need to make decisions on many legitimate requests, said Herbert Pardes, professor of psychiatry and former chief executive officer of New York Presbyterian Hospital. Cost is a critical factor in making these decisions, but it is not the only one. For example, teaching hospitals provide a large amount of care without reimbursement, whether through caring for indigent patients or by paying for promising innovations so that they can be developed to the point of reimbursement.

Hospitals have been deciding how much to invest in new genomic technologies for many years, but the situation is now beginning to change dramatically, according to Pardes. As the cost of sequencing drops and as electronic medical records are integrated into care, many advances are possible. Nevertheless, major questions remain. Who will pay for these advances? How much will they pay for them? How much of an impact will genomics have on improving clinical care? How much information will patients want? Will concerns about privacy be adequately addressed? Pardes also pointed to a practical problem: hospitals have been reluctant to perform genetic testing for inpatients because of the current reimbursement structure. Still, patients and providers want the information that testing can provide.

Data on the comparative effectiveness of genomic testing will take at least a decade to accumulate, Pardes observed, which will delay advances in patient care with this technology. Furthermore, few sources of funds are available to pay for this research, whether from the public or private sectors.

Genomic testing in oncology is an area in which hospitals are deeply involved, and it is rapidly becoming commonplace. Hospitals know that if they do not offer state-of-the-art care, patients and physicians will go elsewhere. Furthermore, building a successful medical genomics program requires gaining experience and developing expertise. Hospitals need to be able to generate, interpret, store, and reinterpret genomic data over time and communicate this information to providers and patients, said Pardes. Genomic information also needs to be integrated into the electronic medical record with appropriate on-demand education and decision support.

Faced with this welter of new information, hospitals have adapted by launching pilot programs. For example, Columbia Presbyterian and Weill Cornell have focused on colorectal cancer, which represents about three-quarters of the gastrointestinal malignancies seen at the hospitals, Pardes said. They are using sequencing technology and bioinformatics to analyze about 50 genes identified as having the greatest potential impact on clinical care. The hospitals are investing about a half-million dollars in the first year in people and equipment to do the sequencing and interpretation in-house. In this way, they will improve inpatient care and learn how best to set up a genomic testing infrastructure and do so in a way that is financially feasible.

Another initiative involves 11 institutions in New York that have come together to create a new genome center. A strong and extensive genomic capacity will be available to all of the scientists within that initiative to understand diseases and improve health, said Pardes. At the same time, the center is designed to contribute to the overall economy of the region.

"Momentum is building with regard to the value of precision medicine," Pardes said. "Ultimately, our intent is to navigate this system by integrating the twin goals of improving health care and reducing cost. Some may think that's impossible; I do not agree with that. I think we can do both."

To realize this potential, Pardes encouraged the academic community to communicate much more extensively with policy makers and politicians. The academic community needs to support research while they also emphasize the need to cut costs without hindering improvements in quality and innovation.

Pardes also encouraged the medical community to involve patients in decisions. Patient involvement can lead to more conservative approaches to procedures. It also helps avoid imposing decisions on patients by outside groups. "When you bring patients into the discussion, good things happen," he said.

ECONOMIC PERSPECTIVES

Health care decisions are not generally based only on cost-effectiveness ratios, Veenstra remarked. A physician offering a test to a patient is trying to produce the greatest benefit for the patient. The health care system offers genetic counseling because of the value it provides for patients. These considerations apply as well in trying to reduce costs in health care. "Health economics isn't about choosing the cheapest thing," said Veenstra. "It's about cutting the things that have the lowest value."

Patient-centered research is focused on what patients need and want to know. They need to know the potential harms and benefits of a test, said Veenstra. They need to understand that a test may produce false positives and false negatives. And they need to have a role in making decisions.

Genomics is probably the most complex area in health care in which to determine value, said Veenstra. The economics of genetic testing is a "morass" and will probably remain so for at least the next two decades.

Grosse observed that cost-effectiveness is one factor in making decisions but not the only one. For example, provider and patient preferences and reimbursement systems also can drive practice. People receive some medical tests more often than cost-effectiveness considerations would dictate, but they receive those tests because patient and provider preferences override other considerations. Even where a test has very little demonstrated clinical utility, a provider may prescribe it, and a patient may agree.

Ramsey pointed out that cost-effectiveness is not a perfect tool for informing decision making, but it is a better tool than the system currently has to determine resource allocation. Most other developed countries have integrated cost-effectiveness analysis into their decision-making processes, and the United States is likely to move in that direction under the Affordable Care Act. For example, accountable care organizations are going to be making decisions about the value of interventions, and they will be promoting high-value interventions and discouraging low-value interventions. As the U.S. health care system seeks to reduce costs while maintaining or improving the quality of care, cost-effectiveness is a tool that can help do that, said Ramsey.

Integrating value into decision making will require getting more stakeholders involved. Stakeholder input is needed to determine what questions to ask and what end points to evaluate, and such input is needed not only at the beginning of a study but throughout the process. "That's the way to move cost-effectiveness from its current state to a state where there's some more acceptance," Ramsey said.

Patients need to demand value, as patient advocacy groups are increasingly doing. These advocacy groups "see families become bankrupt or [living] in severe financial straits being talked into very expensive therapies that actually offer very modest benefit," said Ramsey. Provider groups are starting to understand the costs of their recommendations, though guideline developers have not yet accepted this reality as much as they should. As an example, Ramsey cited treatments for gastrointestinal cancers that provide very similar outcomes yet differ in cost by an order of magnitude.

One option would simply be not to cover, prescribe, or pay for new cancer therapies that offer few benefits, Ramsey observed. Other countries are doing this, but the fragmented delivery system in the United States makes it difficult for all the stakeholders to speak with one voice. Thus, multiple groups will have to understand the benefits of such an approach and advocate for it to bend the curve of rising health care costs.

In response to a question about whether government agencies should seek to place a value on a drug as it progresses through the approval and coverage processes, Ramsey said he did not think that the FDA would get into the business of approving value, "and I'm not even sure that's the right role for them." In addition, Congress has forbidden CMS from making decisions based on QALYs.

Finally, in his reflections on the previous day, Billings concentrated on the role of industry and its view on how value is developed and delivered. Industry does not think about value only for itself, observed Billings. It seeks to provide value to clinicians, payers, and patients when developing new tests and treatments. Nevertheless, clinicians actually represent a very diverse group. They cannot be thought of as a single entity, and no single set of guidelines will apply to every health care provider. Leaving value to be expressed solely in guidelines would be inadequate, he said. Payers are also very heterogeneous in the United States, with some willing to cover molecular diagnostic tests and others not. Industry seeks to deliver value to patients with effective treatments and provides benefit to patients in terms of access to drugs and treatments that might otherwise be unaffordable in return for participating in research. The knowledge gained from this research can then provide further value to patients' families and the community by being a source of information that industry can use to meet patient needs better.

Industry needs some certainty about what value means to these different elements of the health care system, Billings said. Only in that way can it innovate and conduct research to meet that need. At the same time, industry needs the system to be flexible, because its methods are changing rapidly.

A large reference lab may offer between 4,000 and 5,000 kinds of tests, and few of those tests will have gone through rigorous, evidence-based, randomized clinical trials, said Billings. In recent decades, more rigorous, evidence based standards have been developed, particularly for therapeutic interventions. But requiring that randomized clinical trials be conducted for every test is a large burden to put on innovators in very rapidly changing fields, Billings observed, especially when a test targets small patient groups.

A new model is needed for gathering evidence, said Billings. He suggested that electronic medical records systems provide a potentially valuable resource for gathering evidence about value. By combining data from electronic medical records, researchers can study even rare patient groups to determine which variations in a genome are likely to be important for a particular question.

Ramsey said that the burden of collecting evidence that decision makers want for new tests is too high and not feasible for companies. Instead, he suggested coverage with evidence development, which has been piloted at CMS and is now being considered at many health plans. It "would apply very well to this domain," he said. Under coverage with evidence development, insurance companies agree to cover the test with the provision that additional testing is conducted to collect evidence in a structured way. Final approval then depends on whether the evidence demonstrates that the test improves outcomes compared with the standard of care. This model can be risky for test development in that the test may be rejected, which is one reason why manufacturers have been hesitant to pursue this route, said Ramsey. Nevertheless, the time may be right today for manufacturers to place greater reliance on this sort of model.

Ramsey pointed out that coverage with evidence development requires the collection of evidence, but, he said, "I would argue that we have the data." Insurance claims are available that contain information on costs and outcomes, and electronic medical records in the near future will cover most patients. "What we don't have is a mechanism for stitching that information together in real time or in the time that we need to get the data," he said. Building such a mechanism is feasible, though doing so would require examining and resolving a number of issues, including privacy protections.

One participant, however, pointed to the difficulties of collecting evidence under a coverage system in which patients of like circumstances expect similar treatments. "If you want the best evidence development, it needs to be done not through coverage per se but through some kind of an authoritative body making priorities about research," the participant said.

Veenstra agreed with the need to set priorities for research and observed that because gold standard evidence cannot be produced for every test or variant, multiple stakeholders need to come together to identify where this type of evidence is needed. In addition, innovative approaches are needed for gathering evidence, which will require the development of infrastructure and resources to collect that evidence.

He also observed that traditional 1- to 2-year analyses to determine cost-effectiveness are not adequate. More efficient approaches are needed that may be more qualitative than quantitative in nature.

In addition, evidence thresholds are needed to determine at what point patients, physicians, and payers are willing to act on the basis of the evidence in hand, said Veenstra.

ADDITIONAL ISSUES

A participant pointed out that molecular diagnostic testing is being conducted largely in academic medical centers and not in community hospitals, where most cancer patients get care, particularly minority patients. Enhancing the relationships between academic and community hospitals could allow greater access for patients to this type of testing.

Another participant noted that genomics did not create disparities and that it is not going to solve them. The focus should be on whether genomic technologies will exacerbate or ameliorate existing disparities. One issue is the lack of reference genome sequences for underrepresented populations.

Offit observed that genomics can be applied presymptomatically, diagnostically, prognostically, and therapeutically. All these applications are different, and the economic analyses of these applications are different as well.

In addition, a participant pointed out that consumers will increasingly choose the level of insurance coverage they want. The less expensive plans are less likely to cover genomic testing, which could exacerbate disparities.

Another individual pointed to the inevitable involvement of families in genetic testing. Today's health care system is focused on individuals, yet genomic testing often involves individuals and their parents to determine whether a given genetic variant has an effect or not. "The health care system is very poorly organized to deal with families as opposed to individuals," he said. Calonge agreed and noted that this issue brings up additional ones regarding how to perform this testing. Many family members have different insurance companies, which makes coverage of testing difficult even in situations where the medical information may be beneficial. "This is a huge gap that I do not think anyone has quite addressed," Calonge said.

CLOSING REMARKS

Feero observed that a number of research needs had emerged from the workshop (see Box 7-1). He particularly highlighted the need for better methods of measuring and determining the value of patient choice and preference. Practical considerations, such as the ability to afford insurance that includes genomics coverage, need to be accounted for in these measures.

He also noted that better infrastructure is needed to measure economic-related outcomes in addition to traditional measures. Electronic medical records systems need to anticipate the need for the eventual integration of genomic information so that downstream outcomes can be captured.

BOX 7-1
Research Needs Identified by Individual Speakers

In the final session of the workshop, David Veenstra and Scott Ramsey revised and reorganized the workshop themes that W. Gregory Feero presented (see Chapter 1) into four categories of research needs. This box compiles these needs as Veenstra and Ramsey presented them. These themes should not be seen as recommendations of the workshop, but they are promising concepts that warrant further discussion and possible action.

Evidence—Comparative-Effectiveness Research

- Need for development of the evidence base—collaboration, infrastructure with clinical trials groups.
- Need for innovative approaches to the prioritization of comparative-effectiveness research.
- Determining if and how genomic sequence information modifies health care provision and patient outcomes.
- Impact of increasing accuracy of sequencing on patient outcomes and costs.
- Evaluation of proper use of family history to guide medical decision making, integrated into health information technology infrastructure.

Health Economics Methods

- Need better (quicker) approaches and frameworks for performing health economic evaluations of genomic testing.
- Evaluation of evidence thresholds for data in hand versus data that must be obtained, and cost of further research.
- Divergence of economic assessment models in public health, clinical care, and academics.
- In the setting of a disruptive technology and a zero-sum game/shrinking pool of resources, what/who will be replaced, and how will genomic interventions be funded?

continued

BOX 7-1 Continued

Health Economics Applications

- When is genomic sequencing cost-effective? For example, only when it is performed during newborn screening with data being used over the lifespan?
- Better education of genomic scientists regarding economic analysis/integration of economic analysis and ongoing studies.
- Methods/infrastructure (including informatics) in health systems to follow downstream consequences of providing sequence data.
- Is cost reduction demonstrable? Do accountable care organizations provide a possible mechanism for more efficient health care delivery of genomic technologies?
- Study of provider preferences for provision of genomic medicine—evaluation of barriers to implementation.
- Economic incentives for test and evidence development with value-based and specific pricing versus old system (current procedural terminology code stacking).
- Determination of relative contribution of environment/setting on cost-effectiveness.

Patient-Centered Outcomes

- Developing outcomes data on informed consent/study of efficient methods for patient education regarding informed consent.
- Stakeholder engagement; methodology to increase participation in clinical trials.
- Development of improved methods for assessing value/personal utility/patient preference in economic analysis.
- Potential for genomic medicine to exacerbate disparities, including applicability of information to minority populations and socioeconomic status disadvantages. Focus on interventions.

References

ACS (American Cancer Society). 2012. *Cancer facts and figures 2012*. Atlanta, GA: American Cancer Society.
Ashley, E. A., A. J. Butte, M. T. Wheeler, R. Chen, T. E. Klein, F. E. Dewey, J. T. Dudley, K. E. Ormond, A. Pavlovic, A. A. Morgan, D. Pushkarev, N. F. Neff, L. Hudgins, L. Gong, L. M. Hodges, D. S. Berlin, C. F. Thorn, K. Sangkuhl, J. M. Hebert, M. Woon, H. Sagreiya, R. Whaley, J. W. Knowles, M. F. Chou, J. V. Thakuria, A. M. Rosenbaum, A. W. Zaranek, G. M. Church, H. T. Greely, S. R. Quake, and R. B. Altman. 2010. Clinical assessment incorporating a personal genome. *Lancet* 375(9725):1525-1535.
Atherly, A. J., and D. R. Camidge. 2012. The cost-effectiveness of screening lung cancer patients for targeted drug sensitivity markers. *British Journal of Cancer* 106(6):1100-1106.
Balak, M. N., Y. Gong, G. J. Riely, R. Somwar, A. R. Li, M. F. Zakowski, A. Chiang, G. Yang, O. Ouerfelli, M. G. Kris, M. Ladanyi, V. A. Miller, and W. Pao. 2006. Novel D761Y and common secondary T790M mutations in epidermal growth factor receptor-mutant lung adenocarcinomas with acquired resistance to kinase inhibitors. *Clinical Cancer Research* 12(21):6494-6501.
Ball, M. P., J. V. Thakuria, A. W. Zaranek, T. Clegg, A. M. Rosenbaum, X. Wu, M. Angrist, J. Bhak, J. Bobe, M. J. Callow, C. Cano, M. F. Chou, W. K. Chung, S. M. Douglas, P. W. Estep, A. Gore, P. Hulick, A. Labarga, J. H. Lee, J. E. Lunshof, B. C. Kim, J. I. Kim, Z. Li, M. F. Murray, G. B. Nilsen, B. A. Peters, A. M. Raman, H. Y. Rienhoff, K. Robasky, M. T. Wheeler, W. Vandewege, D. B. Vorhaus, J. L. Yang, L. Yang, J. Aach, E. A. Ashley, R. Drmanac, S. J. Kim, J. B. Li, L. Peshkin, C. E. Seidman, J. S. Seo, K. Zhang, H. L. Rehm, and G. M. Church. 2012. A public resource facilitating clinical use of genomes. *Proceedings of the National Academy of Sciences of the United States of America* 109(30):11920-11927.
Basu, A., and D. Meltzer. 2007. Value of information on preference heterogeneity and individualized care. *Medical Decision Making* 27(2):112-127.

Bean, J., G. J. Riely, M. Balak, J. L. Marks, M. Ladanyi, V. A. Miller, and W. Pao. 2008. Acquired resistance to epidermal growth factor receptor kinase inhibitors associated with a novel T854A mutation in a patient with EGFR-mutant lung adenocarcinoma. *Clinical Cancer Research* 14(22):7519-7525.

Beckman, M. G., W. C. Hooper, S. E. Critchley, and T. L. Ortel. 2010. Venous thromboembolism: A public health concern. *American Journal of Preventive Medicine* 38(4 Suppl):S495-S501.

Bradbury, P. A., D. Tu, L. Seymour, P. K. Isogai, L. Zhu, R. Ng, N. Mittmann, M. S. Tsao, W. K. Evans, F. A. Shepherd, N. B. Leighl, and the NCIC Clinical Trials Working Group on Economic Analysis. 2010. Economic analysis: Randomized placebo-controlled clinical trial of erlotinib in advanced non-small-cell lung cancer. *Journal of the National Cancer Institute* 102(5):298-306.

Bristol-Myers Squibb. 2010. Prescribing information for Coumadin (warfarin sodium). Princeton, NJ: Bristol-Myers Squibb. http://www.accessdata.fda.gov/drugsatfda_docs/label/2010/009218s108lbl.pdf (accessed March 19, 2013).

Dentali, F., A. P. Sironi, W. Ageno, S. Turato, C. Bonfanti, F. Frattini, S. Crestani, and M. Franchini. 2012. Non-O blood type is the commonest genetic risk factor for VTE: Results from a meta-analysis of the literature. *Seminars in Thrombosis and Hemostasis* 38(5):535-548.

Dewey, F. E., R. Chen, S. P. Cordero, K. E. Ormond, C. Caleshu, K. J. Karczewski, M. Whirl-Carrillo, M. T. Wheeler, J. T. Dudley, J. K. Byrnes, O. E. Cornejo, J. W. Knowles, M. Woon, K. Sangkuhl, L. Gong, C. F. Thorn, J. M. Hebert, E. Capriotti, S. P. David, A. Pavlovic, A. West, J. V. Thakuria, M. P. Ball, A. W. Zaranek, H. L. Rehm, G. M. Church, J. S. West, C. D. Bustamante, M. Snyder, R. B. Altman, T. E. Klein, A. J. Butte, and E. A. Ashley. 2011. Phased whole-genome genetic risk in a family quartet using a major allele reference sequence. *PLoS Genetics* 7(9):e1002280.

Epstein, R. S., T. P. Moyer, R. E. Aubert, D. J. O'Kane, F. Xia, R. R. Verbrugge, B. F. Gage, and J. R. Teagarden. 2010. Warfarin genotyping reduces hospitalization rates: Results from the MM-WES (Medco-Mayo Warfarin Effectiveness Study). *Journal of the American College of Cardiology* 55(25):2804-2812.

Flowers, C. R., and D. Veenstra. 2004. The role of cost-effectiveness analysis in the era of pharmacogenomics. *Pharmacoeconomics* 22(8):481-493.

Grody, W. W., J. H. Griffin, A. K. Taylor, B. R. Korf, J. A. Heit, and the American College of Medical Genetics Factor V Working Group. 2001. American College of Medical Genetics consensus statement on Factor V Leiden mutation testing. *Genetics in Medicine* 3(2):139-148.

Grosse, S. D., S. Wordsworth, and K. Payne. 2008. Economic methods for valuing the outcomes of genetic testing: Beyond cost-effectiveness analysis. *Genetics in Medicine* 10(9):648-654.

Hampel, H., W. L. Frankel, E. Martin, M. Arnold, K. Khanduja, P. Kuebler, M. Clendenning, K. Sotamaa, T. Prior, J. A. Westman, J. Panescu, D. Fix, J. Lockman, J. LaJeunesse, I. Comeras, and A. de la Chapelle. 2008. Feasibility of screening for Lynch syndrome among patients with colorectal cancer. *Journal of Clinical Oncology* 26(35):5783-5788.

Jick, H., D. Slone, B. Westerholm, W. H. Inman, M. P. Vessey, S. Shapiro, G. P. Lewis, and J. Worcester. 1969. Venous thromboembolic disease and ABO blood type. A cooperative study. *Lancet* 1(7594):539-542.

Kelly, K., J. Crowley, P. A. Bunn, Jr., C. A. Presant, P. K. Grevstad, C. M. Moinpour, S. D. Ramsey, A. J. Wozniak, G. R. Weiss, D. F. Moore, V. K. Israel, R. B. Livingston, and D. R. Gandara. 2001. Randomized phase III trial of paclitaxel plus carboplatin versus vinorelbine plus cisplatin in the treatment of patients with advanced non-small-cell lung cancer: A Southwest Oncology Group trial. *Journal of Clinical Oncology* 19(13):3210-3218.

Kris, M. G., R. B. Natale, R. S. Herbst, T. J. Lynch, Jr., D. Prager, C. P. Belani, J. H. Schiller, K. Kelly, H. Spiridonidis, A. Sandler, K. S. Albain, D. Cella, M. K. Wolf, S. D. Averbuch, J. J. Ochs, and A. C. Kay. 2003. Efficacy of gefitinib, an inhibitor of the epidermal growth factor receptor tyrosine kinase, in symptomatic patients with non-small-cell lung cancer: A randomized trial. *Journal of the American Medical Association* 290(16):2149-2158.

Ladabaum, U., G. Wang, J. Terdiman, A. Blanco, M. Kuppermann, C. R. Boland, J. Ford, E. Elkin, and K. A. Phillips. 2011. Strategies to identify the Lynch syndrome among patients with colorectal cancer: A cost-effectiveness analysis. *Annals of Internal Medicine* 155(2):69-79.

Lockwood, C., G. Wendel, and Committee on Practice Bulletins—Obstetrics. 2011. Practice bulletin no. 124: Inherited thrombophilias in pregnancy. *Obstetrics and Gynecology* 118(3):730-740.

Mallal, S., E. Phillips, G. Carosi, J. M. Molina, C. Workman, J. Tomazic, E. Jagel-Guedes, S. Rugina, O. Kozyrev, J. F. Cid, P. Hay, D. Nolan, S. Hughes, A. Hughes, S. Ryan, N. Fitch, D. Thorborn, A. Benbow, and PREDICT-1 Study Team. 2008. HLA-B*5701 screening for hypersensitivity to abacavir. *New England Journal of Medicine* 358(6):568-579.

Maron, B. J., M. S. Maron, and C. Semsarian. 2012. Genetics of hypertrophic cardiomyopathy after 20 years: Clinical perspectives. *Journal of the American College of Cardiology* 60(8):705-715.

Martin, J. A., B. E. Hamilton, S. J. Ventura, M. J. Osterman, E. C. Wilson, and T. J. Mathews. 2012. *Births: Final data for 2010*. Atlanta, GA: Centers for Disease Control and Prevention.

Medalie, J. H., C. Levene, C. Papier, U. Goldbourt, F. Dreyfuss, D. Oron, H. Neufeld, and E. Riss. 1971. Blood groups, myocardial infarction and angina pectoris among 10,000 adult males. *New England Journal of Medicine* 285(24):1348-1353.

Mok, T. S., Y. L. Wu, S. Thongprasert, C. H. Yang, D. T. Chu, N. Saijo, P. Sunpaweravong, B. Han, B. Margono, Y. Ichinose, Y. Nishiwaki, Y. Ohe, J. J. Yang, B. Chewaskulyong, H. Jiang, E. L. Duffield, C. L. Watkins, A. A. Armour, and M. Fukuoka. 2009. Gefitinib or carboplatin-paclitaxel in pulmonary adenocarcinoma. *New England Journal of Medicine* 361(10):947-957.

Neel, J. V. 1949. The inheritance of sickle cell anemia. *Science* 110(2846):64-66.

Pao, W., V. Miller, M. Zakowski, J. Doherty, K. Politi, I. Sarkaria, B. Singh, R. Heelan, V. Rusch, L. Fulton, E. Mardis, D. Kupfer, R. Wilson, M. Kris, and H. Varmus. 2004. EGF receptor gene mutations are common in lung cancers from "never smokers" and are associated with sensitivity of tumors to gefitinib and erlotinib. *Proceedings of the National Academy of Sciences of the United States of America* 101(36):13306-13311.

Pauling, L., H. A. Itano, S. J. Singer, and I. C. Wells. 1949. Sickle cell anemia: A molecular disease. *Science* 110(2865):543-548.

Plevritis, S. K., A. W. Kurian, B. M. Sigal, B. L. Daniel, D. M. Ikeda, F. E. Stockdale, and A. M. Garber. 2006. Cost-effectiveness of screening BRCA1/2 mutation carriers with breast magnetic resonance imaging. *Journal of the American Medical Association* 295(20):2374-2384.

Ramsey, S. D., D. L. Veenstra, L. P. Garrison, Jr., R. Carlson, P. Billings, J. Carlson, and S. D. Sullivan. 2006. Toward evidence-based assessment for coverage and reimbursement of laboratory-based diagnostic and genetic tests. *American Journal of Managed Care* 12(4):197-202.

Regier, D. A., J. M. Friedman, N. Makela, M. Ryan, and C. A. Marra. 2009. Valuing the benefit of diagnostic testing for genetic causes of idiopathic developmental disability: Willingness to pay from families of affected children. *Clinical Genetics* 75(6):514-521.

Riess, J. W., and H. A. Wakelee. 2012. Metastatic non-small-cell lung cancer management: Novel targets and recent clinical advances. *Clinical Advances in Hematology and Oncology* 10(4):226-234.

Roach, J. C., G. Glusman, A. F. Smit, C. D. Huff, R. Hubley, P. T. Shannon, L. Rowen, K. P. Pant, N. Goodman, M. Bamshad, J. Shendure, R. Drmanac, L. B. Jorde, L. Hood, and D. J. Galas. 2010. Analysis of genetic inheritance in a family quartet by whole-genome sequencing. *Science* 328(5978):636-639.

Russell, R. B., N. S. Green, C. A. Steiner, S. Meikle, J. L. Howse, K. Poschman, T. Dias, L. Potetz, M. J. Davidoff, K. Damus, and J. R. Petrini. 2007. Cost of hospitalization for preterm and low birth weight infants in the United States. *Pediatrics* 120(1):e1-e9.

Schiller, J. H., D. Harrington, C. P. Belani, C. Langer, A. Sandler, J. Krook, J. Zhu, D. H. Johnson, and the Eastern Cooperative Oncology Group. 2002. Comparison of four chemotherapy regimens for advanced non-small-cell lung cancer. *New England Journal of Medicine* 346(2):92-98.

Shah, N. R., and M. B. Bracken. 2000. A systematic review and meta-analysis of prospective studies on the association between maternal cigarette smoking and preterm delivery. *American Journal of Obstetrics and Gynecology* 182(2):465-472.

Shiffman, D., K. Slawsky, L. Fusfeld, J. J. Devlin, and T. F. Goss. 2012. Cost-effectiveness model of use of genetic testing as an aid in assessing the likely benefit of aspirin therapy for primary prevention of cardiovascular disease. *Clinical Therapeutics* 34(6):1387-1394.

Silverstein, M. D., J. A. Heit, D. N. Mohr, T. M. Petterson, W. M. O'Fallon, and L. J. Melton III. 1998. Trends in the incidence of deep vein thrombosis and pulmonary embolism: A 25-year population-based study. *Archives of Internal Medicine* 158(6):585-593.

Sode, B. F., K. H. Allin, M. Dahl, F. Gyntelberg, and B. G. Nordestgaard. 2013. Risk of venous thromboembolism and myocardial infarction associated with Factor V Leiden and prothrombin mutations and blood type. *Canadian Medical Association Journal* 185(5):E229-E237.

Srinivasan, B. S., E. A. Evans, J. Flannick, A. S. Patterson, C. C. Chang, T. Pham, S. Young, A. Kaushal, J. Lee, J. L. Jacobson, and P. Patrizio. 2010. A universal carrier test for the long tail of Mendelian disease. *Reproductive Biomedicine Online* 21(4):537-551.

Vandenbroucke, J. P., J. Rosing, K. W. Bloemenkamp, S. Middeldorp, F. M. Helmerhorst, B. N. Bouma, and F. R. Rosendaal. 2001. Oral contraceptives and the risk of venous thrombosis. *New England Journal of Medicine* 344(20):1527-1535.

Wang, X., B. Zuckerman, C. Pearson, G. Kaufman, C. Chen, G. Wang, T. Niu, P. H. Wise, H. Bauchner, and X. Xu. 2002. Maternal cigarette smoking, metabolic gene polymorphism, and infant birth weight. *Journal of the American Medical Association* 287(2):195-202.

Appendix A

Workshop Agenda

Assessing the Economics of Genomic Medicine: A Workshop
July 17–18, 2012

National Academy of Sciences Building
2101 Constitution Avenue, NW
Washington, DC 20037

WORKSHOP OBJECTIVE

To advance discussions around the clinical implementation of genetic and genomic technologies by examining costs associated with the development and use of genetic and genomic information in the care of individual patients.

Day 1	July 17, 2012
7:45 A.M.–8:30 A.M.	WORKING BREAKFAST
8:30 A.M.–8:35 A.M.	WELCOMING REMARKS

                              Wylie Burke, *Roundtable Co-Chair,*
                              Meeting Moderator
                                  Professor and Chair, Department of Bioethics and Humanities, University of Washington

8:35 A.M.–8:40 A.M.	**CHARGE TO WORKSHOP SPEAKERS AND PARTICIPANTS**
	W. Gregory Feero, *Workshop Chair* Special Advisor to the Director, National Human Genome Research Institute
8:40 A.M.–10:10 A.M.	**SESSION I: ECONOMICS AND GENOMICS**
	Moderators: W. Gregory Feero, National Human Genome Research Institute
	Wylie Burke, University of Washington
8:40 A.M.–9:10 A.M.	**Genomics, Population Health, and Technology**
	James P. Evans Bryson Distinguished Professor of Genetics and Medicine, University of North Carolina at Chapel Hill
9:10 A.M.–9:25 A.M.	**Discussion**
9:25 A.M.–9:55 A.M.	**Intersection of Genomics and Health Economics**
	David L. Veenstra Professor, Pharmaceutical Outcomes Research and Policy Program, University of Washington
9:55 A.M.–10:10 A.M.	**Discussion**
10:10 A.M.–10:25 A.M.	**BREAK**
10:25 A.M.–12:20 P.M.	**SESSION II: PRECONCEPTION CARE AND SEQUENCING**
	Moderators: Robert L. Nussbaum, University of California, San Francisco, School of Medicine
	Wylie Burke, University of Washington

10:25 A.M.–10:30 A.M.	**Case Presentation:** In 2012 a 35-year-old healthy Ashkenazi Jewish female smoker is seen for a preconception visit. Under the current standard care model, targeted carrier status testing is offered. In terms of high effect size variations that would be detected by traditional genetic testing, she is found to be a carrier for Tay-Sachs. In addition, if testing were extended in this scenario beyond what might be considered to be current standard of care, she would be found to harbor a prothrombin gene mutation, as well as variations in CYP2C9 and VKORC, indicating that she is likely to be highly sensitive to warfarin anticoagulation. She is also homozygous for APOE4, but does not have familial hypercholesterolemia. She can be expected to have lower risk variants and variants of unknown significance in accordance with expected population frequencies for the conditions under consideration.
10:30 A.M.–11:00 A.M.	**Stakeholder Presentations (10 minutes each)**

Clinician:

Siobhan M. Dolan
    Associate Professor of Clinical Obstetrics and Gynecology and Women's Health, Albert Einstein College of Medicine

Futurist:

Arthur L. Beaudet
    Henry and Emma Meyer Professor and Chair, Department of Molecular and Human Genetics, Baylor College of Medicine

Patient:

Michelle Gilats
    Genetic Counselor, Chicago Center for Jewish Genetics

11:00 A.M.–11:20 A.M.   Discussion with Speakers and Participants

11:20 A.M.–12:20 P.M.   Economics Panel Discussion

Moderator: David L. Veenstra, University of Washington

Panelists:

Paul R. Billings
  Chief Medical Officer, Life Technologies

Scott Grosse
  Research Economist and Associate Director for Health Services Research and Evaluation, Division of Blood Disorders, National Center on Birth Defects and Developmental Disabilities, Centers for Disease Control and Prevention

Scott Ramsey
  Full Member, Cancer Prevention Program, Division of Public Health Science, Fred Hutchinson Cancer Research Center

12:20 P.M.–1:05 P.M.   **LUNCH**

1:05 P.M.–3:00 P.M.   **SESSION III: UNPROVOKED DVT/ PULMONARY EMBOLISM**

Moderators: Frederick Chen, University of Washington

Wylie Burke, University of Washington

1:05 P.M.–1:10 P.M.   **Case Presentation:** The individual is seen at 40 years of age with progressive left lower extremity swelling and pain. Evaluation reveals an unprovoked deep vein thrombosis in her left lower extremity. She will be treated as an outpatient with low-molecular-weight heparin and warfarin. Targeted testing includes CYP2C9 and VKORC gene analysis.

APPENDIX A   *81*

1:10 P.M.–1:40 P.M.   Stakeholder Presentations (10 minutes each)

                      Clinician:

                      Michael F. Murray
                        Clinical Chief, Genetics Division,
                        Department of Medicine, Brigham and
                        Women's Hospital

                      Futurist:

                      Euan A. Ashley
                        Director, Center for Inherited Cardiovascular
                        Disease, Stanford University School of
                        Medicine

                      Patient:

                      John West
                        Chief Executive Officer, Personalis, Inc.

1:40 P.M.–2:00 P.M.   **Discussion with Speakers and Participants**

2:00 P.M.–3:00 P.M.   **Economics Panel Discussion**

                      Moderator: David L. Veenstra, University of
                      Washington

                      Panelists:

                      Paul R. Billings
                        Chief Medical Officer, Life Technologies

                      Scott Grosse
                        Research Economist and Associate Director
                        for Health Services Research and Evaluation,
                        Division of Blood Disorders, National
                        Center on Birth Defects and Developmental
                        Disabilities, Centers for Disease Control and
                        Prevention

	Scott Ramsey Full Member, Cancer Prevention Program, Division of Public Health Science, Fred Hutchinson Cancer Research Center
3:00 P.M.–3:15 P.M.	**BREAK**
3:15 P.M.–5:10 P.M.	**SESSION IV: CANCER CARE AND SEQUENCING** **Moderators:** Timothy J. Ley, Washington University School of Medicine Wylie Burke, University of Washington
3:15 P.M.–3:20 P.M.	**Case Presentation:** The individual is seen at age 50 with cough, dyspnea, and chest discomfort. Evaluation reveals a lung mass; bronchoscopy and biopsy reveal a non-small-cell lung cancer. Her tumor is found to have variations that allow the use of targeted therapy, and with treatment the patient goes into remission.
3:20 P.M.–3:50 P.M.	**Stakeholder Presentations (10 minutes each)** **Clinician:** Kenneth Offit Chief, Clinical Genetics Service, Department of Human Genetics, Memorial Sloan-Kettering Cancer Center **Futurist:** Stephen Eck Vice President, Global Head of Medical Oncology, Astellas Pharma Global Development

APPENDIX A   83

                            Patient:

                            Richard Heimler
                                  Lung Cancer Survivor and Advocate

3:50 P.M.–4:10 P.M.   **Discussion with Speakers and Participants**

4:10 P.M.–5:10 P.M.   **Economics Panel Discussion**

                            Moderator: David L. Veenstra, University of Washington

                            Panelists:

                            Paul R. Billings
                                  Chief Medical Officer, Life Technologies

                            Scott Grosse
                                  Research Economist and Associate Director for Health Services Research and Evaluation, Division of Blood Disorders, National Center on Birth Defects and Developmental Disabilities, Centers for Disease Control and Prevention

                            Scott Ramsey
                                  Full Member, Cancer Prevention Program, Division of Public Health Science, Fred Hutchinson Cancer Research Center

5:10 P.M.–5:20 P.M.   **WRAP-UP DAY 1**

                            W. Gregory Feero, *Workshop Chair*
                                  Special Advisor to the Director, National Human Genome Research Institute

5:20 P.M.   **ADJOURN DAY 1**

Day 2	July 18, 2012
7:45 A.M.–8:30 A.M.	**WORKING BREAKFAST**
8:30 A.M.–8:35 A.M.	**WELCOMING REMARKS**

    W. Gregory Feero, *Workshop Chair*
        Special Advisor to the Director, National Human Genome Research Institute

8:35 A.M.–11:45 A.M.	Session V: Research Economics
8:35 A.M.–9:00 A.M.	Overview of Major Themes and Research Questions from Day 1

    W. Gregory Feero, *Workshop Chair*
        Special Advisor to the Director, National Human Genome Research Institute

9:00 A.M.–10:00 A.M.	Stakeholder Perspectives on Research Questions

    Moderator: Wylie Burke, University of Washington

    Clinician:

    James P. Evans
        Bryson Distinguished Professor of Genetics and Medicine, University of North Carolina at Chapel Hill

    Researcher:

    Katrina Armstrong
        Professor of Medicine; Associate Director, Abramson Cancer Center; Division Chief, Division of General Internal Medicine, Department of Medicine, University of Pennsylvania School of Medicine

Industry:

Thomas J. White
  Chief Scientific Officer (Retired), Celera Corporation; Regents' Lecturer, University of California, Berkeley

Patient:

Mary Lou Smith
  Cofounder, Research Advocacy Network

Public Health:

Ned Calonge
  President and Chief Executive Officer, The Colorado Trust

Hospital System:

Herbert Pardes
  Professor of Psychiatry and Former Chief Executive Officer, New York–Presbyterian Hospital

10:00 A.M.–10:30 A.M. **Audience Perspectives**

10:30 A.M.–10:45 A.M. **BREAK**

10:45 A.M.–11:45 A.M. **Economics Panel Discussion**

Moderator: David L. Veenstra, University of Washington

Panelists:

Paul R. Billings
  Chief Medical Officer, Life Technologies

Scott Grosse
Research Economist and Associate Director for Health Services Research and Evaluation, Division of Blood Disorders, National Center on Birth Defects and Developmental Disabilities, Centers for Disease Control and Prevention

Scott Ramsey
Full Member, Cancer Prevention Program, Division of Public Health Science, Fred Hutchinson Cancer Research Center

11:45 A.M.–12:00 P.M.   **SESSION VI: WRAP-UP**

11:45 A.M.–12:00 P.M.   **Concluding Remarks**

W. Gregory Feero, *Workshop Chair*
Special Advisor to the Director, National Human Genome Research Institute

12:00 P.M.   **WORKSHOP ADJOURNS**

Appendix B

Speaker Biographical Sketches

Katrina Armstrong, M.D., is professor of medicine, chief of the Division of General Internal Medicine, associate director of the Abramson Cancer Center, and codirector of the Robert Wood Johnson Clinical Scholars Program at the Perelman School of Medicine at the University of Pennsylvania. Dr. Armstrong has completed fellowship training in health services research and clinical epidemiology, as well as clinical training in genetic counseling and testing for breast cancer susceptibility. Her research focuses on the translation of genomic discovery into improvements in cancer control, including understanding the social and economic forces that influence this translation. Her research program has received extensive federal funding, including multiple NIH R01 grants, projects in center grants, and awards from the American Cancer Society, the Department of Defense, and the Robert Wood Johnson Foundation. Dr. Armstrong currently leads two National Cancer Institute–funded centers; they are the Center for Comparative Effectiveness of Genomic Medicine and the Penn Center for Innovation in Personalized Screening funded through the Population-based Research Optimizing Screening through Personalized Regimens initiative.

Euan Angus Ashley, M.R.C.P., D.Phil., FACC, FAHA, is assistant professor in the Division of Cardiovascular Medicine at Stanford University, deputy director of the Stanford Cardiovascular Institute, and director of the Stanford Center for Inherited Cardiovascular Disease. Dr. Ashley graduated with first-class honors in physiology and medicine from the University of Glasgow in 1996. After completing residency at the John Radcliffe Hospital of the University of Oxford, he joined the Ph.D. program in molecular cardiology. His work elucidating a role for intramyocardial nitric oxide in cardiac contractility attracted Young Investigator awards from the Medical Research Society of the United Kingdom, the European Society of Cardiology, and the American Heart Association. In 2002 he moved to California to join the Cardiology Division of Stanford University, first as a fellow and later as faculty. His laboratory is focused on the molecular genetics of inherited cardiovascular disease. In 2008 the team was awarded the National Innovation Award from the American Heart Association, and in 2009 Dr. Ashley was awarded the NIH Director's New Innovator award. In 2010 he led the team that carried out a clinical interpretation of a whole human genome, and in 2011 the team extended the approach to a family of four. Dr. Ashley is part of the Myocardial Applied Genomics Network, a member of the leadership group of the American Heart Association's Council on Functional Genomics, and codirector of the NIH-funded Research Training Program in Myocardial Biology at Stanford.

Arthur L. Beaudet, M.D., received a B.S. in biology magna cum laude from College of the Holy Cross in 1963 and his M.D. cum laude from Yale University in 1967. He then did 2 years of pediatrics residency at Johns Hopkins Hospital and spent 2 years as a research associate at the National Institutes of Health in Bethesda before going to Baylor College of Medicine (BCM) in 1971, where he has remained to the present. Dr. Beaudet has published more than 250 original research articles on diverse aspects of mammalian genetics. His contributions included the demonstration of mutations in cultured somatic cells in the 1970s at a time when such evidence was still considered novel. He published extensively on inborn errors of metabolism, particularly on urea cycle disorders. In 1988 his group was the first to describe uniparental disomy in humans. He has had long-standing interests in somatic gene therapy and in cystic fibrosis. He has studied genomic imprinting as it relates to Prader-Willi and Angelman syndromes, including identification of the causative role of the UBE3A gene in Angelman syndrome and of the importance of the snoRNA genes in the pathogenesis of Prader-Willi syndrome. In 2004 Beaudet and a BCM team of investigators were the first in the United States to introduce array comparative genomic hybridization into the clinical lab, and they have gone on to play a leadership role in the transformative impact of this technology on clinical genetics. This work has led to a focus on the role of the CHRNA7 neuronal nicotinic receptor gene in mental retardation, autism, and schizophrenia. Dr. Beaudet is one of the editors of the *Metabolic and Molecular Bases of Inherited Disease* textbook (6th through 8th and electronic editions), and he has served on numerous editorial boards and national review panels. He was president of the American Society of Human Genetics in 1998 and is an elected member of the Association of American Physicians and the Institute of Medicine of the National Academy of Sciences. Dr. Beaudet is currently the Henry and Emma Meyer Distinguished Service Professor and chair of the Department of Molecular and Human Genetics at BCM and Texas Children's Hospital in Houston.

Paul R. Billings, M.D., Ph.D., is a board-certified internist and clinical geneticist who serves as chief medical officer of Life Technologies Corporation, where he seeks to improve patient care through expanding the use of medically relevant genomic technologies in clinical settings. Dr. Billings brings extensive expertise and health care experience to the areas of genomics and molecular medicine. Most recently, he served as director and chief scientific officer of the Genomic Medicine Institute at El Camino Hospital, the largest community hospital in the Silicon Valley. He has been a member of the Secretary's Advisory Committee on Genetics, Health and Society at the U.S. Department of Health and Human Services and currently serves on the Scientific Advisory Board of the U.S. Food and Drug Administration and the Genomic Medicine Advisory Committee at the Department of Veterans Affairs. He has been a founder or chief executive officer of companies involved in genetic and diagnostic medicine, including GeneSage, Omicia, and CELLective Dx Corporation.

Wylie Burke, M.D., Ph.D., is professor and chair of the Department of Bioethics and Humanities at the University of Washington. She received a Ph.D. in genetics and an M.D. from the University of Washington and completed a residency in internal medicine at the University of Washington. She was a medical genetics fellow at the University of Washington from 1981 to 1982. From 1983 to 2000, Dr. Burke was a member of the Department of Medicine at the University of Washington, where she served as associate director of the internal medicine residency program and founding director of the University of Washington's Women's Health Care Center. She was appointed chair of the Department of Medical History and Ethics (now the Department of Bioethics and Humanities) in October 2000. She is also an adjunct professor of medicine and epidemiology and a member of the Fred Hutchinson Cancer Research Center. She is a member of the Institute of Medicine and the Association of American Physicians and is a past president of the American Society of Human Genetics. Dr. Burke's research addresses the social, ethical, and policy implications of genetics, including responsible conduct of genetic and genomic research, genetic test evaluation, and implications of genomic health care for underserved populations. She is director of the University of Washington Center for Genomics and Healthcare Equality, a National Human Genome Research Institute Center of Excellence in Ethical, Legal, and Social Implications research, and codirector of the Northwest-Alaska Pharmacogenomics Research Network.

Ned Calonge, M.D., is the president and chief executive officer of The Colorado Trust, which was established in 1985. The Colorado Trust works closely with nonprofit organizations across the state to improve health and well-being. Before joining The Colorado Trust in 2010, Dr. Calonge served as the chief medical officer of the Colorado Department of Public Health and Environment. He also served as the chief of the Department of Preventive Medicine for the Colorado Permanente Medical Group and was a family physician for 10 years. His current academic appointments include serving as associate professor of family medicine, Department of Family Medicine, University of Colorado Denver (UCD) School of Medicine, and associate professor of preventive medicine and biometrics, UCD Colorado School of Public Health. Nationally, Dr. Calonge is the immediate past chair of the U.S. Preventive Services Task Force and a member of the Task Force on Community Preventive Services for the Centers for Disease Control and Prevention (CDC). He is chair of the CDC's Evaluating Genomic Applications for Practice and Prevention workgroup and is a consultant for and past member of the Secretary's Advisory Committee on Heritable Disorders in Newborns and Children. Dr. Calonge earned a B.A. in chemistry from Colorado College, an M.P.H. from the University of Washington, and an M.D. from the University of Colorado. He was elected to the Institute of Medicine in 2011.

Frederick Chen, M.D., M.P.H., is chief of family medicine at Harborview Medical Center and associate professor in the Department of Family Medicine at the University of Washington, where he teaches health policy, conducts research, and sees patients. He attended medical school at the University of California, San Francisco, and received his master's of public health in epidemiology from the University of California, Berkeley. After completing his residency in family medicine at the University of Washington, Dr. Chen was a Robert Wood Johnson Clinical Scholar, where he developed his research interest in health policy and medical education. He then moved to Washington, DC, as the Kerr White Scholar at the U.S. Agency for Healthcare Research and Quality. At the University of Washington, he has been the lead faculty for the WWAMI Underserved Pathway, medical director for the Washington State Patient-Centered Medical Home Collaborative, and a researcher in the Rural Health Research Center. He is the medical director for the Washington State employees' health plan and served as the chair of the American Academy of Family Physicians' Subcommittee on Genomics.

Siobhan Dolan, M.D., M.P.H., is an associate professor in the Department of Obstetrics and Gynecology and Women's Health at the Albert Einstein College of Medicine/Montefiore Medical Center in New York City. She currently serves as a medical advisor to March of Dimes and is on the faculty of the human genetics program at Sarah Lawrence College, where she teaches public health genetics and genomics. Dr. Dolan is board certified in both obstetrics and gynecology and clinical genetics. She received a master's degree in public health from Columbia University. Dr. Dolan maintains her clinical practice in the Bronx, where she provides prenatal care to women using an innovative group model, and she serves as an attending physician in the Division of Reproductive Genetics at Montefiore. Her research interests focus on the integration of genetics into maternal and child health, specifically looking at ways to apply advances in genetics and genomics to prevent birth defects and preterm birth.

Stephen L. Eck, M.D., Ph.D., is vice president and global head of oncology medical science at Astellas Pharma Global Development. He is directly responsible for the oversight of oncology drug development plans. Much of this work is focused on special cancer populations for which unique biology enables the development of personalized cancer therapies. From 2007 to 2011, Dr. Eck served as vice president of translational medicine and pharmacogenomics at Eli Lilly and Company, where he was responsible for the clinical pharmacology components of drug development, including both early-phase clinical studies and late-stage drug development studies. His group also developed the biomarkers and companion diagnostics needed for effective decision making and for tailoring therapeutics to the right patient population. Before joining Eli Lilly, Dr. Eck served in a variety of drug development leadership roles at Pfizer Inc. Dr. Eck is a board-certified hematologist/oncologist with broad drug development experience in oncology and neuroscience. He is a fellow of the American Association for the Advancement of Science (Pharmaceutical Sciences). He serves on the Scientific Advisory Board of the Alliance for Cancer Gene Therapy Foundation and is a member of the executive committee of the Fairbanks Institute, an institution dedicated to developing data banks to enable personalized medicine. He also serves on the advisory board of the Keck Graduate School and is a board member of the Personalized Medicine Coalition.

James P. Evans, Ph.D., M.D., is the Bryson Distinguished Professor of Genetics and Medicine at the University of North Carolina at Chapel Hill. He directs adult and cancer genetics services and serves as editor-in-chief of *Genetics in Medicine*, the official journal of the American College of Medical Genetics. After obtaining his M.D. and Ph.D. from the University of Kansas, Dr. Evans served as resident and chief resident of internal medicine at the University of North Carolina and then trained in medical genetics at the University of Washington in Seattle. He is board certified in internal medicine, medical genetics, and molecular diagnostics. He remains clinically active in both genetics and general medicine. Dr. Evans' research interests focus on cancer genetics, pharmacogenomics, the use of next-generation genomic analytic technologies in medicine, and broad issues of how genetic information is used and perceived. He has been extensively involved in policy issues related to genetics and medicine and has published widely on these topics. He was an advisor to the U.S. Secretary of Health and Human Services on the subject of genetics, health, and society from 2004 to 2010 and is actively involved both nationally and internationally in the education of high court judges about genetic and scientific matters. In 2010, Dr. Evans testified before Congress regarding the regulation of direct-to-consumer genetic testing and advised the Government Accountability Office on the same subject. In 2011, he addressed the U.S. Presidential Commission on Bioethics regarding genetic testing.

W. Gregory Feero, M.D., Ph.D., is special advisor to the director for genomic medicine at the National Human Genome Research Institute. Dr. Feero obtained his M.D./Ph.D. from the University of Pittsburgh School of Medicine with a Ph.D. in human genetics. He then completed his residency in family medicine at the Maine–Dartmouth Family Medicine Residency Program in Augusta, Maine, where he still sees patients. He is an associate professor in the Department of Community and Family Medicine at Dartmouth Medical School. Dr. Feero is board certified in family medicine and holds licenses in Maine and West Virginia. He has written numerous peer-reviewed and invited publications.

Michelle Gilats, M.S., is a licensed genetic counselor at the Ann and Robert H. Lurie Children's Hospital of Chicago. She received her B.S. from the University of Wisconsin, Madison, and her M.S. in genetic counseling from the University of California, Berkeley. She has worked in the areas of pediatric, prenatal, and cancer counseling. At Lurie Children's Hospital she helps coordinate the pediatric neurofibromatosis clinic. For the Center for Jewish Genetics, Ms. Gilats provides education regarding Jewish genetic disorders and hereditary cancers to community groups and counsels individuals and couples about carrier testing. She also answers inquiries from people across the country and internationally. She helps coordinate the center's screening programs by interacting with participants prior to registration to ensure that testing is appropriate for that individual or couple. She relays screening results to all participants and provides follow-up testing and counseling as needed.

Scott Grosse, Ph.D., is research economist and associate director for Health Services Research and Evaluation in the Division of Blood Disorders, National Center on Birth Defects and Developmental Disabilities, CDC, in Atlanta, Georgia. He has degrees in economics and public health from the University of Michigan. Dr. Grosse conducts research on the health outcomes and economic benefits of the early identification of hereditary conditions and prevention of preventable conditions that manifest in early childhood. He has written and cowritten more than 150 journal articles and book chapters. In addition to research on specific diseases, he publishes policy analyses and cost-effectiveness analyses of public health strategies, such as newborn screening and genetic testing. Dr. Grosse also publishes on health economic measures and methods. These include summary health measures, methods of assessing the economic value of diagnosis or prevention, the history of the $50,000 per QALY threshold for cost-effectiveness, and human capital measures of productivity losses.

Richard Heimler is a former nonprofit executive who was diagnosed with non-small-cell lung cancer at the age of 44 in 2004. In the past 8 years he has been diagnosed six times with malignant tumors in the lung, brain, and thorax. After Mr. Heimler's cancer progressed to Stage 4, his tissue was tested and found to be positive for ALK. He subsequently enrolled in a clinical trial of Xalkori, during which his tumors have shrunk, his pulmonary function has increased, and his overall health has improved. Since his initial diagnosis, Mr. Heimler has been an active lung cancer advocate, providing inspiration and advice to newly diagnosed lung cancer patients and raising awareness among the media, politicians, and the general public so that he and fellow lung cancer survivors may look forward to celebrating many more birthdays.

Timothy J. Ley, M.D., received his M.D. from Washington University Medical School in St. Louis in 1978 and performed his internal medicine residency at Massachusetts General Hospital. He completed fellowships in hematology and oncology at the NIH and at Washington University and joined the faculty at Washington University in 1986. He now holds the Lewis T. and Rosalind B. Apple Chair in Oncology, is professor of medicine and of genetics at Washington University, and serves as an associate director of the Genome Institute (for Cancer Genomics). Dr. Ley is a past president of the American Society for Clinical Investigation and is a fellow of the American Association for the Advancement of Science and the American Academy of Arts and Sciences and a member of the Institute of Medicine. He has performed pioneering studies of acute myeloid leukemia genomes and modeled several key AML mutations in the mouse.

Michael F. Murray, M.D., is a former primary care provider and is now the clinical chief of genetics at Brigham and Women's Hospital in Boston. He trained in internal medicine at the Cleveland Clinic and then went on to do fellowships in infectious diseases and medical genetics. Dr. Murray directs the annual course in The Genetic Basis of Adult Medicine: What the Primary Care Provider Needs to Know, as well as directing a combined residency training program in internal medicine and medical genetics. He leads the Adult Genetics Clinic at Brigham and Women's Hospital, where more than 400 patients per year are evaluated, diagnosed, and treated. He is one of the principal investigators in an NHGRI project titled "Integration of Whole Genome Sequencing into Clinical Medicine." His research interests include the integration of electronic family health history tools into medical practice and the use of whole genome testing in medicine.

Robert L. Nussbaum, M.D., is chief of the Division of Medical Genetics in the Department of Medicine and a faculty member in the Institute of Human Genetics at the University of California, San Francisco. He focuses on three main areas of research: (1) an investigation of the genetic contribution to Parkinson's disease; (2) a long-standing effort to understand the rare X-linked disease known as the oculocerebrorenal syndrome of Lowe, characterized by congenital cataracts, Fanconi syndrome of the renal proximal tubules, neurological dysfunction, and developmental delay; and (3) a translational research effort to assess the value of "personalized medicine," the application of genetic and genomic approaches to improving patient care. Dr. Nussbaum seeks to evaluate if and how genetic and genomic information about an individual can be used effectively to improve health care by improving outcomes, reducing adverse reactions, lowering costs, and promoting health through risk education. Dr. Nussbaum is seeking to develop collaborative research efforts with clinician-researchers interested in studying how applying genomics can improve patient care.

Kenneth Offit, M.D., is chief of the clinical genetics service at Memorial Sloan-Kettering Cancer Center and a professor of medicine and public health at the Weill College of Medicine at Cornell University. His research group first described and characterized the most common BRCA2 mutation associated with breast and ovarian cancer, was among the first to measure prospectively the impact of preventive ovarian surgery in individuals carrying BRCA mutations, and performed the first genome-wide association study of BRCA2 breast cancer. His lab is currently defining genomic markers of risk for breast, colon, and prostate cancer and lymphoma. Dr. Offit has received a career research recognition award from the American Cancer Society and is a member of the Board of Scientific Counselors of the U.S. National Cancer Institute.

Herbert Pardes, M.D., former president and chief executive officer of New York–Presbyterian Hospital and New York–Presbyterian Healthcare

System, is executive vice chairman of the board of New York–Presbyterian Hospital. Nationally recognized for his broad expertise in education, research, clinical care, and health policy, he is an ardent advocate of academic medical centers, humanistic care, and the power of technology and innovation to transform 21st-century medicine. Under his leadership, New York–Presbyterian has become one of the most highly regarded and comprehensive health care institutions in the world. The hospital is top-ranked in the New York metropolitan area and is consistently ranked among the best academic medical institutions in the nation, according to *U.S. News & World Report*. Before joining the hospital in 1999, Dr. Pardes served as vice president for health sciences at Columbia University and dean of the faculty of medicine at Columbia University College of Physicians and Surgeons. A noted psychiatrist, he served as director of the National Institute of Mental Health and U.S. assistant surgeon general during the Carter and Reagan administrations and was president of the American Psychiatric Association. He received his medical degree from the State University of New York in Brooklyn and completed his residency in psychiatry at Kings County Hospital in Brooklyn, with additional psychoanalytic training at the New York Psychoanalytic Institute.

Scott Ramsey, M.D., Ph.D., is a full member in the Cancer Prevention Program at the Fred Hutchinson Cancer Research Center, where he directs the Research and Economic Assessment in Cancer and Healthcare group, a multidisciplinary team devoted to clinical and economic evaluations of new and existing cancer prevention, screening, and treatment technologies. He is also a professor in the School of Medicine, School of Pharmacy, and Institute for Public Health Genetics at the University of Washington. Trained in medicine and economics, his primary research interest is studying the economic aspects of new medical technologies. Dr. Ramsey is a leader in the field of comparative-effectiveness research. He is past president of the International Society of Pharmacoeconomics and Outcomes Research and has served on the National Cancer Policy Forum of the Institute of Medicine.

Mary Lou Smith, J.D., M.B.A., is a cofounder of the Research Advocacy Network. She is a two-time breast cancer survivor and serves as cochair of the Patient Representative Committee of the Eastern Cooperative Oncology Group and the Patient Advocate Committee of the Radiation Therapy Oncology Group. Ms. Smith also serves on the National Comprehensive Cancer Network Breast Cancer Screening and Treatment Guidelines Committees and the Translational Breast Cancer Research Consortium. In addition, she serves on the advocate core of the Department of Defense Center of Excellence for Individualization of Therapy for Breast Cancer and on the advocate core of the Komen Promise Grant at Indiana University.

Ms. Smith was a community member of Chicago's Rush University Medical Center institutional review board for 10 years. She is past president of the Y-ME National Breast Cancer Organization and has served on the Cancer Leadership Council and the National Breast Cancer Coalition's board of directors. Ms. Smith was involved in the development of numerous managed care products for the Blue Cross and Blue Shield Association, including a pediatric cancer network. She has a juris doctorate with a health law certification and a master's degree in business administration.

David L. Veenstra, Pharm.D., Ph.D., is a professor in the Pharmaceutical Outcomes Research and Policy Program in the Department of Pharmacy and a member of the Institute for Public Health Genetics at the University of Washington in Seattle. He graduated from the University of California, San Francisco, with doctoral degrees in clinical pharmacy and computational chemistry. He conducted his postdoctoral training in outcomes research with the University of Washington, including a 1-year externship with Roche Global Pharmacoeconomics. Dr. Veenstra's primary research interests are the clinical, economic, and policy implications of using genomic information in health care. His other major research interest is the development of simulation models for chronic diseases. Dr. Veenstra's major research projects include evaluation of warfarin pharmacogenomics and decision modeling in breast and lung cancer to inform research prioritization and stakeholder decision making. His research is funded through grants from the Centers for Disease Control and Prevention, National Cancer Institute, National Human Genome Research Institute, and National Institute for General Medical Sciences. He has worked extensively with the Academy of Managed Care Pharmacy to develop guidelines and train decision makers in the practical application of cost-effectiveness models. Dr. Veenstra is a member of the Evaluation of Genomic Applications in Practice and Prevention working group and an author or coauthor of 100 peer-reviewed publications and 5 book chapters.

John West, M.B.A., M.S., is chief executive officer of Personalis, Inc. He was first involved in DNA sequencing, and DNA sequence interpretation, starting in 1982. In the 1980s he led the development of an automated DNA sequencing system based on pattern recognition from autoradiographs, and he licensed software from the lab of Roger Staden at the Medical Research Council, Cambridge, United Kingdom, for sequence assembly and analysis. In the 1990s Mr. West was general manager and subsequently president of Princeton Instruments, a company focused on low-light scientific imaging used in fluorescent-automated DNA sequencing. In 2001 Mr. West joined Applied Biosystems as vice president of genetic analysis. He was subsequently promoted to vice president, DNA platforms. In 2004 Mr. West became chief executive officer of Solexa Ltd., a venture capital–backed UK company focused primarily on single-molecule DNA sequencing. In 2005 he led Solexa's reverse merger into U.S.-based Lynx Therapeutics. The company introduced its first system in mid-2006. Mr. West negotiated the January 2007 acquisition of Solexa by Illumina, Inc., and stayed as vice president of the DNA sequencing business there into 2008. From 2009 through mid-2011 Mr. West served as chief executive officer of ViaCyte, Inc., a company leveraging stem cell technology to develop a diabetes cell therapy. In mid-2009 Illumina introduced its individual genome sequencing service, and Mr. West and his family were the first family of four sequenced by the company. Mr. West received his B.S. and M.S. engineering degrees from MIT and earned an M.B.A. from the University of Pennsylvania's Wharton School.

Thomas J. White, Ph.D., received his B.A. in chemistry from Johns Hopkins University and his Ph.D. in biochemistry from the University of California, Berkeley. His postdoctoral research was carried out at the University of California, San Francisco, Medical Center and at the University of Wisconsin, Madison. From 1978 to 1989 Dr. White held the positions of vice president of research and associate director of research and development at the biotechnology firm Cetus Corporation. He worked on the discovery, research, and development of human proteins as therapeutics, such as Betaseron for the treatment of multiple sclerosis and interleukin-2 for renal cell carcinoma, and on research, forensic, and diagnostic products using polymerase chain reaction (PCR) technology. From 1989 to 2000 he worked for Roche Molecular Systems, a diagnostics division of Hoffmann–La Roche. As senior vice president of research and development, he was responsible for La Roche's research and development on PCR-based tests on the Cobas instrument systems for diagnosing infectious diseases, genetic diseases, and cancer; for screening the blood supply for human immunodeficiency virus, hepatitis C virus, and hepatitis B virus; and for developing new applications of PCR for basic research, forensics, and the human genome project. From 2001 to 2011 Dr. White was chief scientific officer at Celera Corporation. He retired in June 2011 and is the regents' lecturer at the University of California, Berkeley, for 2012–2013.

Appendix C

Statement of Task

An ad hoc planning committee will plan and conduct a public workshop that will assess the potential economic impact that the advent of genomic medicine may have on clinical practice and research. The workshop will feature presentations and discussions from an array of stakeholders which may include health economists, providers, payers, guideline developers, patients, and regulators. The goal of the workshop will be to advance discussions around the clinical implementation of genetic and genomic technologies by examining costs associated with the development and use of genetic and genomic information in the care of individual patients. The planning committee will develop the workshop agenda, select and invite speakers and discussants, and moderate the discussions. An individually authored summary of the workshop will be prepared by a designated rapporteur in accordance with institutional policy and procedures.

Appendix D

Registered Attendees

Adetayo Adewolu
Prince George's County Health Department

Kris Anderson
Freelance

Katrina Armstrong
University of Pennsylvania

Naomi Aronson
Blue Cross and Blue Shield Association

Euan Ashley
Stanford University

Eric Assaraf
WRG

Carlos Avila
Abt Associates

Erin Balogh
Institute of Medicine

Arthur Beaudet
Baylor College of Medicine

Paul Billings
Life Technologies

Gregory Bloss
National Institute on Alcohol Abuse and Alcoholism

Bruce Blumberg
Kaiser Permanente

Juli Bollinger
Johns Hopkins University

Denise Bonds
National Heart, Lung, and Blood Institute

Vence Bonham
National Human Genome Research Institute

Joann Boughman
American Society of Human Genetics

Jen Bowman
American Clinical Laboratory Association

Pamela Bradley
American Association for Cancer Research

Kenneth Brigham
Emory University

Stacye Bruckbauer
FasterCures

Wylie Burke
University of Washington

Ned Calonge
The Colorado Trust

Sarah Carter
J. Craig Venter Institute

C. Thomas Caskey
Baylor College of Medicine

Lon Castle
Express Scripts

Frederick Chen
University of Washington

Melina Cimler
Illumina

Erin Cole
Mayo Clinic

Richard Conroy
National Institutes of Health

Sara Copeland
Health Resources and Services Administration

Cecilia Copperman
Genetic Alliance

Jeff Cossman
United States Diagnostics Standards, Inc.

Claude Desjardins
Johns Hopkins University

Patricia Deverka
Center for Medical Technology Policy

Noel Doheny
Epigenomics, Inc.

Siobhan Dolan
Albert Einstein College of Medicine/Montefiore Medical Center

Maria DeTolve Donoghue
G&M Consulting Services

Subash Duggirala
Centers for Medicare & Medicaid Services

Victor Dzau
Duke University Health System

APPENDIX D

Peggy Eastman
Oncology Times

Stephen Eck
Astellas

Matt Elrod
American Physical Therapy Association

Rebecca English
Institute of Medicine

Raith Erickson
Complete Genomics

James Evans
University of North Carolina at Chapel Hill

W. Gregory Feero
National Human Genome Research Institute

J. Michael Fitzmaurice
Agency for Healthcare Research and Quality

Heather Tollerson Flannery
Obesity PPM

Mark Fleury
American Association for Cancer Research

Andrew Freedman
National Cancer Institute

Michelle Gilats
Chicago Center for Jewish Genetics

Geoffrey Ginsburg
Duke University, Institute for Genome Sciences and Policy

Pamela Goetz
National Coalition for Cancer Survivorship

Jeanne Gorman
JMG Associates

Scott Grosse
Centers for Disease Control and Prevention

John Haaga
National Institute on Aging

Evan Hadley
National Institute on Aging

Kelly Haenlein
Genentech

Chris Havasy
Presidential Commission for the Study of Bioethical Issues

Kristy Hawley
American Medical Association

Richard Heimler
Patient

C. J. Hoban
Multiple Myeloma Research Foundation

India Hook-Barnard
National Academy of Sciences

Gillian Hooker
National Human Genome Research Institute

Kathi Huddleston
Inova Translational Medicine Institute

Jean Jenkins
National Human Genome Research Institute

Janet Jenkins-Showalter
Roche/Genentech

Irene Jillson
Georgetown University

Heajin Jung
Heajin Jung Law Firm

Francis Kalush
U.S. Food and Drug Administration

Jeffrey Kant
University of Pittsburgh Medical Center

Sharon Kardia
University of Michigan

David Kaufman
Genetics and Public Policy Center

Rebecca Kelly
American College of Cardiology

Mohamed Khan
British Columbia Cancer Agency

Muin Khoury
Centers for Disease Control and Prevention

Roger Klein
University of South Florida School of Medicine

John Lauerman
Bloomberg News

Thomas Lehner
National Institute of Mental Health

Jennifer Leib
HealthFutures, LLC

Debra Leonard
Weill Cornell Medical College

Laura Levit
Institute of Medicine

Timothy Ley
Washington University

Jeffrey Lin
Johns Hopkins University Applied Physics Laboratory

Stephen Lincoln
InVitae/Genomic Health

Nicole Littmann
Quorum Consulting

Michele Lloyd-Puryear
Office of Rare Diseases Research, National Institutes of Health

Jenny Luray
BD

Julie Lynch
U.S. Department of Veterans Affairs

Elizabeth Mansfield
Office of In Vitro Diagnostics and Radiological Health, U.S. Food and Drug Administration

Gary Marchant
Arizona State University

Saralyn Mark
NASA

Priscilla Markwood
Association of Pathology Chairs

Robert McCormack
Veridex, LLC

Scott McGoohan
American Clinical Laboratory Association

Kathryn McLaughlin
Health Resources and Services Administration

Brian McTigue
Partners Healthcare

Kala Menon
Booz Allen Hamilton

Douglas C. Monroe
Kaiser Permanente

Jennifer Moser
U.S. Department of Veterans Affairs

Sharon Murphy
Institute of Medicine

Marc Newman
TeleCenter

Robert Nussbaum
University of California, San Francisco

Kenneth Offit
Memorial Sloan-Kettering Cancer Center

Lydia Pan
Pfizer Inc.

Herbert Pardes
New York–Presbyterian Hospital

Meeta Patnaik
Transtek Clinical Systems, Inc.

Michelle Penny
Eli Lilly and Company

Aidan Power
Pfizer Inc.

Victoria Pratt
Quest Diagnostics

Oscar Puig
Roche

Bruce Quinn
Foley Hoag LLP

Scott Ramsey
Fred Hutchinson Cancer Research Center

Kate Reed
National Coalition for Health Professional Education in Genetics

John Reppas
Neurotechnology Industry Organization

Steven Richardson
Genomic Health, Inc.

Denise Robinson
bioTheranostics, Inc.

Allen Roses
Duke University

Julie Sakowski
University of California, San Francisco

Carol Sardinha
CAS Healthcare Associates

Mari Savickis
American Medical Association

Derek Scholes
National Human Genome Research Institute

Joan Scott
National Coalition for Health Professional Education in Genetics

Cecili Sessions
Air Force Medical Support Agency

Paul Sheives
Biotechnology Industry Organization

Leah Silva
Genetic Alliance

Tania Simoncelli
Office of Medical Products and Tobacco, U.S. Food and Drug Administration

Naoko Simonds
National Cancer Institute

Mary Lou Smith
Research Advocacy Network

James Sorace
U.S. Department of Health and Human Services

Noemie Sportiche
U.S. Department of Health and Human Services

Tamara Stuchlak
Air Force Medical Support Agency

Jayson Swanson
Genetic Alliance

Katie Johansen Taber
American Medical Association

Sharon Terry
Genetic Alliance

Tamar Thompson
Kimbell & Associates

Kuo Tong
QUORUM Consulting

David Veenstra
University of Washington

Xiaobin Wang
Johns Hopkins Bloomberg School of Public Health

Michael Watson
American College of Medical Genetics and Genomics

Meredith Weaver
American College of Medical Genetics and Genomics

Lale White
XIFIN, Inc.

Ross White
The Hastings Center

Catherine Wicklund
Northwestern University

Jacqueline Wieneke
U.S. Food and Drug Administration

David Wierz
OCI, LLC

Wendy Wifler
Agendia Inc.

Mara Wilber
Genetic Alliance

Erin Wilhelm
Georgetown University

Marc Williams
Geisinger Health System

Mary Williams
Association for Molecular Pathology

Stephen Williams
Marwood Group

Martin Willie
Universal Genomic Systems, Inc.

Rina Wolf
XIFIN, Inc.

Marta Wosinska
U.S. Food and Drug Administration